Texas STAAR Grade 3 Math Made Easy

Study Guide with Key Concepts, Review & Practice

Dr. A. Nazari

Grade 3 Math Study Guide

Hi there, math learner! 

This study guide has everything you need to learn Grade 3 math. Each topic explains the **key idea**, shows you **examples**, and gives you **practice problems** to try.

Ready? Let's go! 

 How to Use This Book 📘

💡 Key Concept

The main idea explained clearly. Read this first to understand the topic.

✏️ Examples

See how to solve problems step by step. Follow along carefully!

🏋️ Practice

Try problems on your own! Answers are in the back of the book.

⭐ Owl Tips

Look for Owl's tips — they help you remember important tricks!

Tip: Work through one topic at a time. Don't rush — understanding is more important than speed!

$\mathbf{X^1}$ Math Symbols You Should Know $\mathbf{X^1}$

You'll see these symbols throughout this book!

Symbol	Name	What It Means	
$+$	Plus (Add)	Put numbers together.	$3 + 5 = 8$
$-$	Minus (Subtract)	Take away from a number.	$9 - 4 = 5$
$\times$	Times (Multiply)	Add equal groups.	$4 \times 3 = 12$
$\div$	Divide	Split into equal groups.	$12 \div 3 = 4$
$=$	Equals	Both sides are the same.	$2 + 3 = 5$
$>$	Greater Than	The left number is bigger.	$7 > 3$
$<$	Less Than	The left number is smaller.	$2 < 9$
$\frac{1}{2}$	Fraction Bar	Part of a whole.	$\frac{1}{2}$ means 1 out of 2 equal parts

66 *Remember the Alligator!* **99**

*The **greater than** ($>$) and **less than** ($<$) symbols are like an alligator's mouth. The alligator always wants to eat the **bigger** number!*

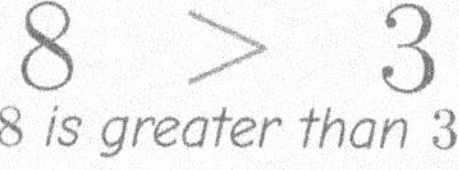

$$8 > 3 \qquad\qquad 2 < 9$$

8 is greater than 3   2 is less than 9

Key Math Words for Grade 3

- **Sum** — the answer when you add
- **Difference** — the answer when you subtract
- **Product** — the answer when you multiply
- **Quotient** — the answer when you divide
- **Factor** — a number you multiply
- **Array** — objects in rows and columns
- **Fraction** — a part of a whole

- **Numerator** — the top number in a fraction
- **Denominator** — the bottom number
- **Equation** — a math sentence with $=$
- **Estimate** — a smart guess, close to the real answer
- **Perimeter** — the distance around a shape
- **Area** — the space inside a shape
- **Rounding** — making a number simpler by going to the nearest ten or hundred

Find more at
ViewMath.com/Grade3

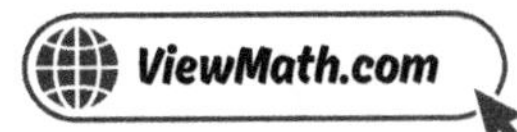

Multiplication Table

Use this chart to practice your multiplication facts!

×	1	2	3	4	5	6	7	8	9	10	11
1	1	2	3	4	5	6	7	8	9	10	11
2	2	4	6	8	10	12	14	16	18	20	22
3	3	6	9	12	15	18	21	24	27	30	33
4	4	8	12	16	20	24	28	32	36	40	44
5	5	10	15	20	25	30	35	40	45	50	55
6	6	12	18	24	30	36	42	48	54	60	66
7	7	14	21	28	35	42	49	56	63	70	77
8	8	16	24	32	40	48	56	64	72	80	88
9	9	18	27	36	45	54	63	72	81	90	99
10	10	20	30	40	50	60	70	80	90	100	110
11	11	22	33	44	55	66	77	88	99	110	121

How to Use This Table

To find **4 × 7**:

1. Find **4** in the left column (blue).
2. Find **7** in the top row (blue).
3. Follow the row and column until they meet: the answer is **28**!

 Table of Contents

Here's what we'll explore together!

 Let's learn and have fun!

1

Number Sense & Place Value

What's Inside

 1.1 Place Value: Ones, Tens, Hundreds 

 In this lesson you will learn:

- Understand that a digit's value depends on its position
- Identify the ones, tens, and hundreds places
- Write numbers in expanded form

What Is Place Value?

Place value tells us how much a digit is worth based on where it sits in a number.

Look at the number **527**:

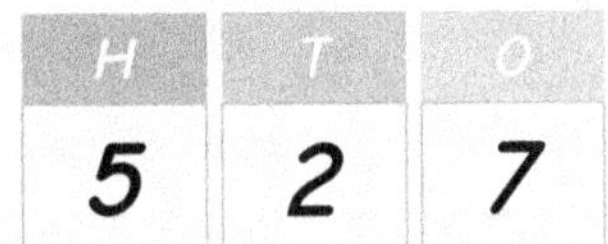

- The **5** is in the **hundreds** place $\rightarrow 500$
- The **2** is in the **tens** place $\rightarrow 20$
- The **7** is in the **ones** place $\rightarrow 7$

Expanded form: $527 = 500 + 20 + 7$

A **0** is a placeholder. In 308, the 0 means zero tens: $308 = 300 + 0 + 8$.

66 Think of place value like floors in a building! Ones are on the first floor, tens on the second, hundreds on the top. Higher floor = bigger value! 99

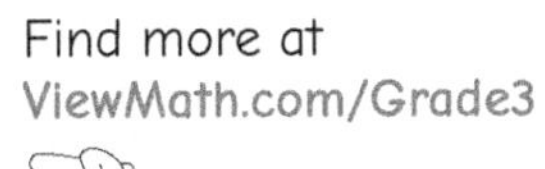

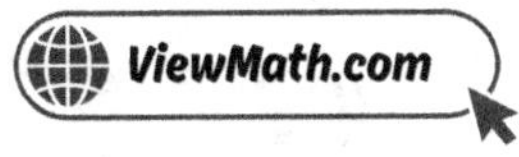

Show 364 with Base-Ten Blocks

1. 3 hundreds = 300

2. 6 tens = 60

3. 4 ones = 4

> ✓ **Answer:** $364 = 300 + 60 + 4$

What Is the Value of 6 in 461?

1. Find which place the 6 is in: **tens**

2. The tens place means we multiply by 10

> ✓ **Answer:** The value of 6 in 461 is 60.

✏ Place Value Practice ✏

1. What is the value of <u>3</u> in 382? ___________

2. What is the value of <u>6</u> in 561? ___________

3. Write 259 in expanded form: _______________________________

4. Write the number for $600 + 30 + 9$: ___________

5. Write 406 in expanded form: _______________________________

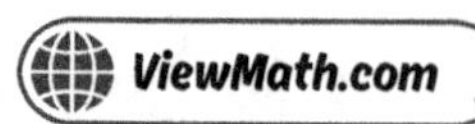

6. Lily has 3 hundred-dollar bills, 5 ten-dollar bills, and 2 one-dollar bills. How much money does Lily have?

Answer: _________________ dollars

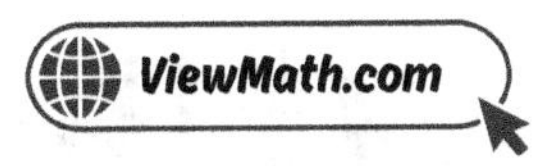

★ 1.2 Place Value: Thousands ★

◎ In this lesson you will learn:

- *Extend place value to the thousands place and understand the $\times 10$ relationship between places*
- *Read and write 4-digit numbers in expanded form*
- *Compare and order 4-digit numbers using $<$, $>$, and $=$*

🎓 Welcome to the Thousands!

The **thousands** place is the fourth digit from the right. Each place is **10 times** the place to its right:

$$1 \xrightarrow{\times 10} 10 \xrightarrow{\times 10} 100 \xrightarrow{\times 10} 1{,}000$$

Look at **3,472**:

Thousands	Hundreds	Tens	Ones
3	4	7	2
3,000	400	70	2

$3{,}472 = 3{,}000 + 400 + 70 + 2$

Use a **comma** to separate the thousands: write 3,472, not 3472.

Zeros are placeholders: $4{,}059 = 4{,}000 + 0 + 50 + 9$.

Comparing 4-digit numbers: Start from the **leftmost** digit. Compare the thousands first, then hundreds, then tens, then ones. The first place where the digits differ tells you which number is greater.

 **66** *Each place is 10 times bigger than the one to its right! So 3 thousands = 30 hundreds = 300 tens = 3,000 ones.* **99**

Expand 6,835

1. *6 is in the thousands place:* 6,000

2. *8 is in the hundreds place:* 800

3. *3 is in the tens place:* 30

4. *5 is in the ones place:* 5

Answer: $6,835 \;=\; 6,000 + 800 + 30 + 5$

Compare 4,738 and 4,692

1. *Compare the* **thousands**: *both are 4 — they match.*

2. *Compare the* **hundreds**: $7 > 6$.

3. *Since the hundreds digit is greater, 4,738 is the larger number.*

Answer: $4,738 \;>\; 4,692$

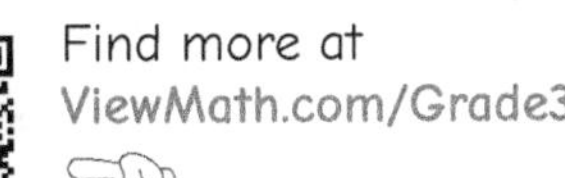
Find more at
ViewMath.com/Grade3

Place Value and Comparing

1. Write 5,206 in expanded form. _______________________

2. What is the value of the digit 9 in 9,314? _______________

3. Write the number for $7,000 + 400 + 60 + 1.$ _______________

4. Compare using <, >, or =:   2,856 _______ 2,912

5. Compare using <, >, or =:   6,450 _______ 6,405

6. Order from least to greatest: 3,210,   3,120,   3,201.

 1.3 Comparing & Ordering Numbers

◎ In this lesson you will learn:

- Compare numbers using $<$, $>$, and $=$
- Order numbers from least to greatest and greatest to least
- Use place value to compare multi-digit numbers

 Comparing Numbers

We use special symbols to compare numbers:

Symbol	Meaning	Example
$>$	greater than	$8 > 3$
$<$	less than	$2 < 9$
$=$	equal to	$5 = 5$

How to compare:

1. The number with **more digits** is greater. $(1{,}234 > 987)$
2. Same number of digits? Compare from the **left**.
3. Find the first digit that is **different** — it decides!

Ordering means putting numbers in a line from smallest to biggest (or biggest to smallest).

66 *The symbol is like a hungry alligator — its mouth always opens toward the BIGGER number!*
$8 > 3$ 99

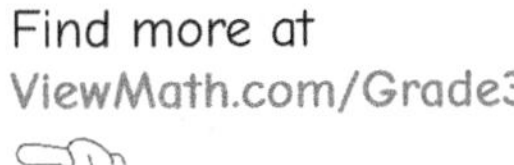
Find more at
ViewMath.com/Grade3

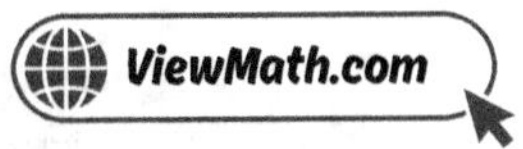

Compare 637 and 682

Both are 3-digit numbers, so compare digit by digit from the left:

1. Hundreds: $6 = 6$ (same — move right)

2. Tens: $3 < 8$ (3 is less than 8)

> ✓ **Answer:** 637 $<$ 682

Order 429, 387, 451 from Least to Greatest

1. Compare hundreds: 387 has 3, the others have 4. So 387 is the smallest.

2. Compare 429 and 451: tens are 2 vs. 5. Since $2 < 5$, $429 < 451$.

> ✓ **Answer:** $387,$ $429,$ 451

✏ Comparing & Ordering Practice ✏

Compare: Write $<$, $>$, or $=$

1. 345 _____ 398

2. 721 _____ 712

3. 500 _____ 500

4. $3,409$ _____ $3,490$

Order from Least to Greatest

5. 284,   197,   342 $\longrightarrow$ _______________________________________

6. 5,120,   5,021,   5,210 $\longrightarrow$ _______________________________________

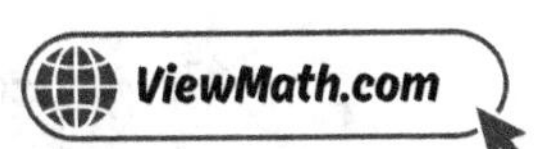

⭐ 1.4 Rounding to the Nearest 10 or 100 ⭐

◎ In this lesson you will learn:

- Round numbers to the nearest 10
- Round numbers to the nearest 100
- Use the "5-or-more" rule

🎓 What Is Rounding?

Rounding means replacing a number with a nearby "friendly" number that is easier to work with.

Rounding to the Nearest 10:

1. Look at the **ones** digit.
2. 0–4: round **down** (keep the tens digit).
3. 5–9: round **up** (add 1 to the tens digit).
4. Change the ones digit to 0.

Rounding to the Nearest 100:

1. Look at the **tens** digit.
2. 0–4: round **down** (keep the hundreds digit).
3. 5–9: round **up** (add 1 to the hundreds digit).
4. Change the tens and ones digits to 0.

❝ 5 or more? Go up!   4 or less? Stay low! **❞**

Round 73 to the Nearest 10

1. The ones digit is **3**.

2. $3 < 5$, so round **down**.

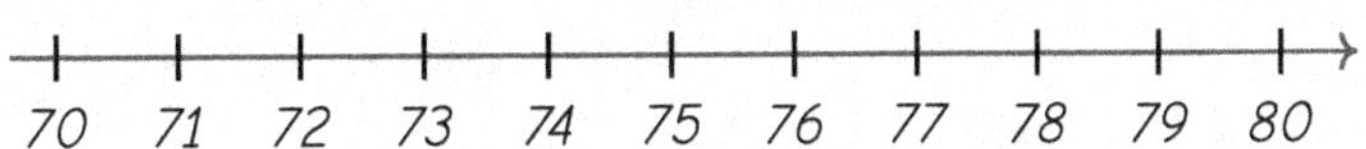

73 is closer to 70 than to 80.

> ✓ **Answer:** $73 \approx 70$

Round 678 to the Nearest 100

1. The tens digit is **7**.

2. $7 \geq 5$, so round **up**.

3. 678 rounds up to 700.

> ✓ **Answer:** $678 \approx 700$

✏ Rounding Practice ✏

Round to the Nearest 10

1. $34 \approx$ _____________

2. $67 \approx$ _____________

3. $85 \approx$ _____________

Round to the Nearest 100

4. $341 \approx$ _____________

5. $550 \approx$ _____________

6. $782 \approx$ _____________

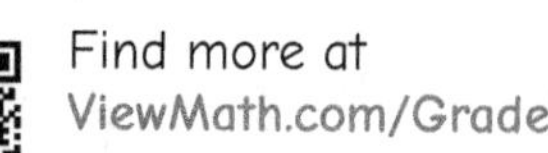

Find more at
ViewMath.com/Grade3

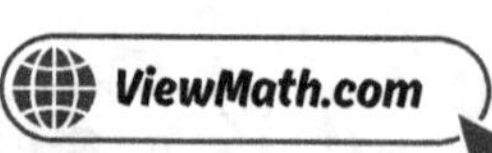

★ 1.5 Even & Odd Numbers ★

◎ In this lesson you will learn:

- Tell if a number is even or odd
- Understand why even numbers split into 2 equal groups
- Know the addition rules for even and odd

🎓 Even and Odd Numbers

Even numbers can be split into 2 equal groups with nothing left over.

$$2, \quad 4, \quad 6, \quad 8, \quad 10, \quad 12, \quad 14, \quad \ldots$$

Odd numbers have 1 left over when split into 2 groups.

$$1, \quad 3, \quad 5, \quad 7, \quad 9, \quad 11, \quad 13, \quad \ldots$$

Quick check: *just look at the* **last digit** *(the ones place).*

- *Ends in* $0, 2, 4, 6, 8 \rightarrow$ **even**
- *Ends in* $1, 3, 5, 7, 9 \rightarrow$ **odd**

Addition rules:

Even + Even	=	Even
Odd + Odd	=	Even
Even + Odd	=	Odd

Find more at
ViewMath.com/Grade3

> *Every even number is a result of multiplying something by 2! That's why you can always split it into 2 equal groups.*

Is 347 Even or Odd?

1. Look at the ones digit: **7**

2. 7 is in the odd list $(1, 3, 5, 7, 9)$

> ✓ **Answer:** 347 *is odd.*

Is 1,286 Even or Odd?

1. Look at the ones digit: **6**

2. 6 is in the even list $(0, 2, 4, 6, 8)$

> ✓ **Answer:** 1,286 *is even.*

✏ Even & Odd Practice ✏

Even or Odd? Write E or O.

1. 14 _______

2. 27 _______

3. 130 _______

4. *Is 5,403 even or odd?* _______________

5. *What is* $8 + 6$*?* _________   *Is the answer even or odd?* ___________

6. *What is* $7 + 4$*?* _________   *Is the answer even or odd?* ___________

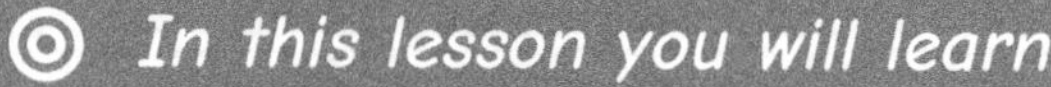

◎ In this lesson you will learn:

- Understand the **ten-thousands** place and read 5-digit numbers
- Write numbers up to 10,000 in expanded form

🎓 The Ten-Thousands Place

You know the ones, tens, hundreds, and thousands places. The next place is the **ten-thousands** place — the 5th digit from the right.

$10,000 = 10$ groups of $1,000$. That's ten thousand!

Ten-Thousands	Thousands	Hundreds	Tens	Ones
1	0	0	0	0
10,000	0	0	0	0

Each place is $10\times$ bigger than the place to its right:

$$1 \xrightarrow{\times 10} 10 \xrightarrow{\times 10} 100 \xrightarrow{\times 10} 1,000 \xrightarrow{\times 10} 10,000$$

The number right before $10,000$ is $9,999$. So $9,999 + 1 = 10,000$.

> ❝ Remember the pattern! Each place is 10 times bigger than the one to its right. The ten-thousands place is $10 \times 1,000 = 10,000$! ❞

Expand 10,000

1. 1 is in the ten-thousands place: 10,000

2. All other digits are 0

> ✅ **Answer:** $10,000 \ = \ 10,000 + 0 + 0 + 0 + 0$

What Is the Value of Each Digit in 10,000?

Ten-Th	Th	H	T	O
1	0	0	0	0

1. The digit 1 is in the ten-thousands place

2. Its value is 10,000

3. We read it as "ten thousand"

> ✅ **Answer:** The 1 in 10,000 has a value of 10,000.

✏️ Ten-Thousands Practice ✏️

1. How many thousands are in 10,000? __________   __________

2. What digit is in the ten-thousands place of 10,000?

____

3. What number is 1,000 more than 9,000?

4. What number is 1 more than 9,999? __________

5. 2,000, 4,000, 6,000, ________, ________

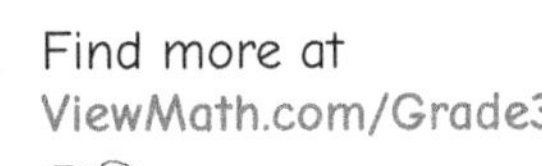

Find more at
ViewMath.com/Grade3

6. *A school library has* 10,000 *books. How many groups of* 1,000 *is that?*

Answer: _________________ *groups*

 1.7 Place Value: Hundred-Thousands 

◎ In this lesson you will learn:

- Understand the **hundred-thousands** place — the 6th digit from the right
- Read, write, and expand numbers up to 100,000

🎓 The Hundred-Thousands Place

You know up to the ten-thousands place. Now let's go one more step — the **hundred-thousands** place!

$100,000 = 100$ groups of $1,000$ — or 10 groups of $10,000$.

Here is a place value chart showing all 6 places:

HunTh	TenTh	Th	H	T	O
1	0	0	0	0	0
100,000	0	0	0	0	0

The pattern continues: $10,000 \times 10 = 100,000$. Every time we multiply by 10, we move one place to the left!

Look at **53,472**:

HunTh	TenTh	Th	H	T	O
0	5	3	4	7	2
0	50,000	3,000	400	70	2

$53,472 = 50,000 + 3,000 + 400 + 70 + 2$

  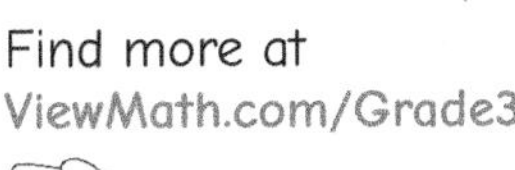
Find more at
ViewMath.com/Grade3

> **❝** *Use commas to separate groups of three digits from the right. This makes big numbers easy to read: 53,472 instead of 53472.* **❞**

Expand 72,385

1. *7 is in the ten-thousands place:* 70,000

2. *2 is in the thousands place:* 2,000

3. *3 is in the hundreds place:* 300

4. *8 is in the tens place:* 80

5. *5 is in the ones place:* 5

> ✔ **Answer:** $72{,}385 \;=\; 70{,}000 + 2{,}000 + 300 + 80 + 5$

How Do We Say 100,000?

1. *100,000 has 6 digits*

2. *The 1 is in the hundred-thousands place*

3. *We say: "one hundred thousand"*

> ✔ **Answer:** $100{,}000$ **= one hundred thousand**

Find more at
ViewMath.com/Grade3

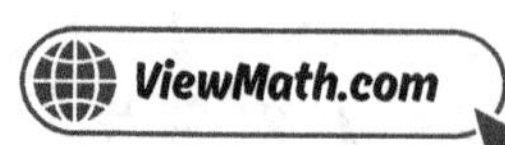

Hundred-Thousands Practice

1. What is the value of <u>6</u> in 62,415? ___________   ___________

2. What is the value of <u>8</u> in 38,701? ___________   4. How many groups of 1,000 make 100,000?

3. How many groups of 10,000 make 100,000?   ___________

5. Write 45,908 in expanded form: _______________________________________

6. A stadium holds 60,000 fans. What digit is in the ten-thousands place, and what is its value?

Answer: _________________

⭐ 1.8 Place Value: To Millions ⭐

◎ In this lesson you will learn:

- *Read, write, and expand numbers up to 999,999*
- *Compare large numbers digit by digit from the left*
- *Know that 999,999 + 1 = 1,000,000 (one million)*

🎓 All Six Places Together

Now you know every place from ones up to hundred-thousands. Let's see them all!

Look at the number **472,856**:

HunTh	TenTh	Th	H	T	O
4	7	2	8	5	6
400,000	70,000	2,000	800	50	6

Expanded form: $472,856 = 400,000 + 70,000 + 2,000 + 800 + 50 + 6$

Standard form: $472,856$ (the usual way we write it)

Comparing large numbers: *Start from the* **left** *and compare digit by digit. The first place where digits differ tells you which number is bigger.*

Example: $347,200$ *vs.* $352,100$ — *both start with 3, but the ten-thousands digits are 4 vs. 5, so* $347,200 < 352,100$.

What comes after $999,999$? *The next number is* **1,000,000** — *one million!*

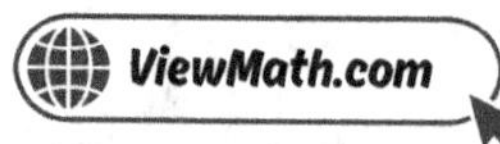

> **ⓘ TIP** Use the **comma trick**: put a comma after every three digits from the right. This separates the
> thousands period from the ones period and makes big numbers easy to read.

Expand 305,419

1. 3 *hundred-thousands* = 300,000

2. 0 *ten-thousands* = 0

3. 5 *thousands* = 5,000

4. 4 *hundreds* = 400

5. 1 *ten* = 10

6. 9 *ones* = 9

> ✅ **Answer:** 305,419 = 300,000 + 5,000 + 400 + 10 + 9

Compare 456,000 *and* 465,000

1. Hundred-thousands: both are 4 — same!

2. Ten-thousands: 5 vs. 6 — 6 is bigger!

> ✅ **Answer:** 456,000 < 465,000

Find more at
ViewMath.com/Grade3

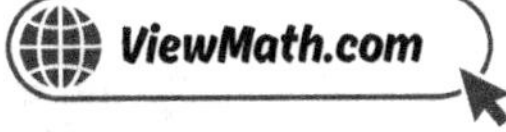

Big Numbers Practice

1. What is the value of <u>5</u> in 514,823? _______________

2. What is the value of <u>8</u> in 283,041? _______________

3. Write 256,374 in expanded form: ____

4. Write the standard form for $700,000 + 40,000 + 3,000 + 200 + 10 + 8$: _______________

5. Compare — write >, <, or =:   892,100 _______ 829,100

6. What number is 1 more than 999,999?

Answer: _______________

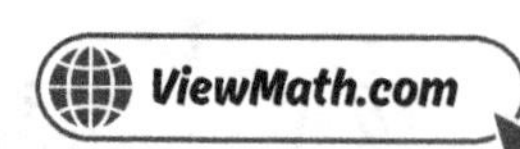

2

Addition & Subtraction

What's Inside

★ 2.1 Adding 3-Digit Numbers ★

◎ In this lesson you will learn:

- Add 3-digit numbers with and without regrouping
- Use place value to line up digits correctly

🎓 How to Add 3-Digit Numbers

To add big numbers, line them up by **place value** and add one column at a time, starting from the **right** (ones).

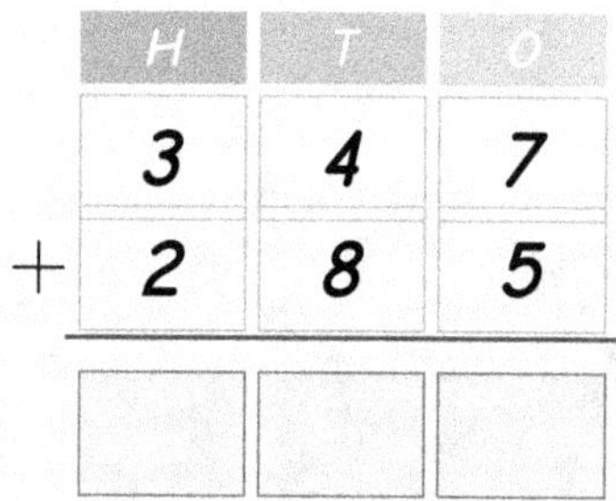

1. Add the **ones** column first.
2. Then the **tens** column.
3. Then the **hundreds** column.

If any column adds up to 10 or more, we **regroup** (carry): write the ones digit below the line and carry the tens digit to the next column.

Example: Ones: $7 + 5 = 12$. Write 2, carry 1 to tens.

> **❝** Always start from the **right**! And if a column adds to 10 or more, don't forget to carry the extra digit to the next column. **❞**

Find more at
ViewMath.com/Grade3

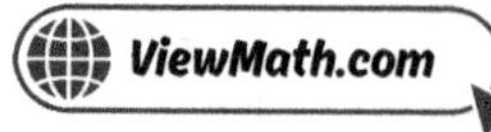

Add $234 + 152$ (No Regrouping)

$$
\begin{array}{r}
2\ \ 3\ \ 4 \\
+\ 1\ \ 5\ \ 2 \\
\hline
\end{array}
$$

1. Ones: $4 + 2 = 6$

2. Tens: $3 + 5 = 8$

3. Hundreds: $2 + 1 = 3$

> ✔ **Answer:** $234 + 152 = 386$

Add $347 + 285$ (With Regrouping)

$$
\begin{array}{r}
{}^{1}\ \ {}^{1}\ \ \ \\
3\ \ 4\ \ 7 \\
+\ 2\ \ 8\ \ 5 \\
\hline
6\ \ 3\ \ 2
\end{array}
$$

1. **Ones:** $7 + 5 = 12$. Write 2, carry 1 to tens.

2. **Tens:** $4 + 8 + 1 = 13$. Write 3, carry 1 to hundreds.

3. **Hundreds:** $3 + 2 + 1 = 6$.

> ✔ **Answer:** $347 + 285 = 632$

  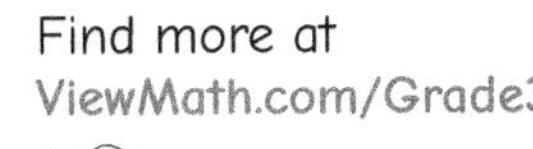

Find more at
ViewMath.com/Grade3

ViewMath.com

3-Digit Addition Practice

1. $312 + 245 =$ _______________

2. $421 + 136 =$ _______________

3. $356 + 478 =$ _______________

4. $189 + 247 =$ _______________

5. $567 + 285 =$ _______________

6. A bakery made 375 muffins in the morning and 248 muffins in the afternoon. How many muffins were made in total?

Answer: _______________ muffins

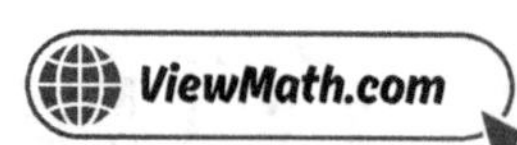

★ 2.2 Subtracting 3-Digit Numbers ★

◎ In this lesson you will learn:

- Subtract 3-digit numbers with and without regrouping
- Know when and how to borrow across columns

🎓 How to Subtract 3-Digit Numbers

Just like addition, line up the digits by **place value** and work from **right to left**.

But here's the twist: if the top digit is smaller than the bottom digit, we need to **regroup** (borrow).

Borrowing means taking 1 from the column to the left. That 1 becomes 10 in the current column.

Steps:

1. Subtract the **ones**. If the top digit is smaller, borrow 1 ten (= 10 ones) from tens.
2. Subtract the **tens**. If the top digit is smaller, borrow 1 hundred (= 10 tens) from hundreds.
3. Subtract the **hundreds**.

Watch out for zeros! If the tens digit is 0 (like in 600), you can't borrow from it. Go to the hundreds first, then borrow down.

> 66 When you borrow, the digit you take from goes **down by** 1, and the digit you give to goes up by 10! 99

Find more at
ViewMath.com/Grade3

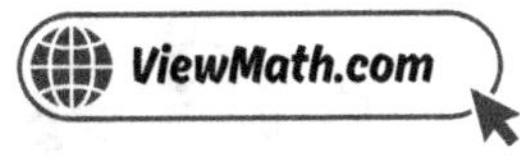

Subtract $586 - 243$ *(No Regrouping)*

$$\begin{array}{r} 5\ \ 8\ \ 6 \\ -\ 2\ \ 4\ \ 3 \\ \hline \end{array}$$

1. *Ones:* $6 - 3 = 3$
2. *Tens:* $8 - 4 = 4$
3. *Hundreds:* $5 - 2 = 3$

✓ **Answer:** $586 - 243 = 343$

Subtract $532 - 275$ *(With Regrouping)*

$$\begin{array}{r} 5\ \ 3\ \ 2 \\ -\ 2\ \ 7\ \ 5 \\ \hline 2\ \ 5\ \ 7 \end{array}$$

1. **Ones:** $2 < 5$. Can't subtract! Borrow 1 ten. Now: $12 - 5 = 7$.
2. **Tens:** The 3 became 2 (we borrowed). $2 < 7$. Borrow 1 hundred! Now: $12 - 7 = 5$.
3. **Hundreds:** The 5 became 4. $4 - 2 = 2$.

✓ **Answer:** $532 - 275 = 257$

Find more at
ViewMath.com/Grade3

ViewMath.com

3-Digit Subtraction Practice

1. $897 - 453 =$ _______________

2. $675 - 234 =$ _______________

3. $524 - 367 =$ _______________

4. $843 - 578 =$ _______________

5. $601 - 258 =$ _______________

6. A school has 750 students. If 382 are girls, how many are boys?

Answer: _______________ boys

Find more at
ViewMath.com/Grade3

★ 2.3 Adding 4-Digit Numbers ★

◎ In this lesson you will learn:

- Extend the column addition method to 4-digit numbers
- Regroup across ones, tens, hundreds, and thousands

🎓 Adding 4-Digit Numbers

Adding 4-digit numbers works exactly like 3-digit addition — just add one more column on the left: the **thousands!**

Th	H	T	O
2	3	4	5
+ 1	4	2	3
3	7	6	8

1. Start from the **ones** (right side).
2. Add each column, moving left.
3. **Regroup** whenever a column totals 10 or more.
4. Don't forget the **thousands** column at the end!

A **0** in any column just means "add nothing" for that spot. For example, in 3,057, the hundreds digit is 0.

> 66 If you can add 3-digit numbers, you can add 4-digit numbers! It's just one extra column. You've got this! 99

Find more at
ViewMath.com/Grade3

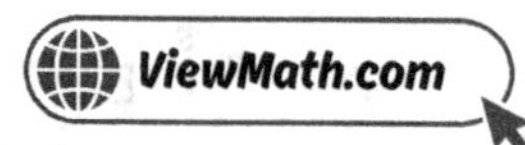

Add $3,214 + 2,563$ *(No Regrouping)*

$$\begin{array}{r} 3,214 \\ + \ 2,563 \\ \hline \end{array}$$

1. *Ones:* $4 + 3 = 7$
2. *Tens:* $1 + 6 = 7$
3. *Hundreds:* $2 + 5 = 7$
4. *Thousands:* $3 + 2 = 5$

✓ Answer: $3,214 + 2,563 = 5,777$

Add $4,685 + 2,738$ *(With Regrouping)*

$$\begin{array}{cccc} 1 & 1 & 1 & \\ 4 & 6 & 8 & 5 \\ + \ 2 & 7 & 3 & 8 \\ \hline 7 & 4 & 2 & 3 \end{array}$$

1. **Ones:** $5 + 8 = 13.$ *Write 3, carry 1.*
2. **Tens:** $8 + 3 + 1 = 12.$ *Write 2, carry 1.*
3. **Hundreds:** $6 + 7 + 1 = 14.$ *Write 4, carry 1.*
4. **Thousands:** $4 + 2 + 1 = 7.$

✓ Answer: $4,685 + 2,738 = 7,423$

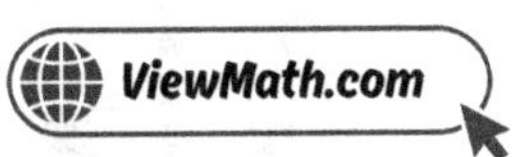

4-Digit Addition Practice

1. $1,234 + 3,452 =$ ______________

2. $2,103 + 4,561 =$ ______________

3. $3,456 + 2,789 =$ ______________

4. $1,875 + 4,368 =$ ______________

5. $5,697 + 2,845 =$ ______________

6. Town A has 4,356 people. Town B has 3,789 people. How many people live in both towns combined?

Answer: ______________ people

Get Online

Find more at
ViewMath.com/Grade3

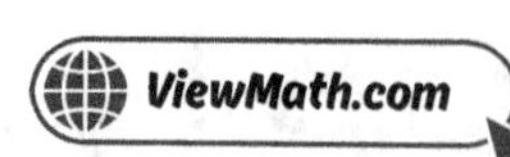

★ 2.4 Subtracting 4-Digit Numbers ★

◎ In this lesson you will learn:

- Subtract 4-digit numbers with and without regrouping
- Handle borrowing across zeros in larger numbers

🎓 Subtracting 4-Digit Numbers

Subtracting 4-digit numbers works just like 3-digit subtraction — you just have one more column: the **thousands**!

Thousands ← Hundreds ← Tens ← Ones (start here!)

1. Start from the **ones** (right side).
2. Subtract each column, moving left.
3. If the top digit is **smaller** than the bottom digit, **borrow** 1 from the column to the left.
4. When a column is **0** and you need to borrow, keep going left until you find a digit to borrow from!

Check your work: Add the answer to the smaller number. If you get the bigger number back, you're right!

❝ Zeros can look scary, but just keep borrowing from the next column to the left. One step at a time! ❞

Find more at
ViewMath.com/Grade3

Subtract $7,869 - 2,435$

$$\begin{array}{r} 7,869 \\ -\,2,435 \\ \hline 5,434 \end{array}$$

1. Ones: $9 - 5 = 4$

2. Tens: $6 - 3 = 3$

3. Hundreds: $8 - 4 = 4$

4. Thousands: $7 - 2 = 5$

✓ **Answer:** $7,869 - 2,435 = 5,434$

Subtract $4,000 - 1,637$

All those zeros mean we need to borrow through each column!

$$\begin{array}{r} 4,000 \\ -\,1,637 \\ \hline 2,363 \end{array}$$

1. **Ones:** $0 < 7$. Borrow from tens — but tens is 0!

2. Borrow from hundreds — also 0! Borrow from thousands: $4 \rightarrow 3$.

3. Now hundreds becomes 10, borrow to tens: $10 \rightarrow 9$, tens gets 10.

4. Borrow from tens to ones: $10 \rightarrow 9$, ones gets 10.

5. Ones: $10 - 7 = 3$. Tens: $9 - 3 = 6$. Hundreds: $9 - 6 = 3$. Thousands: $3 - 1 = 2$.

✓ **Answer:** $4,000 - 1,637 = 2,363$

Find more at
ViewMath.com/Grade3

4-Digit Subtraction Practice

1. 8,976 − 3,542 = _____________

2. 5,234 − 2,678 = _____________

3. 7,401 − 3,856 = _____________

4. 6,005 − 2,738 = _____________

5. 5,000 − 2,346 = _____________

6. A store had 6,250 items. They sold 3,487 items. How many items are left?

Answer: _____________ items

⭐ 2.5 Estimating Sums & Differences ⭐

◎ In this lesson you will learn:

- Estimate sums and differences by rounding to the nearest 10 or 100
- Use estimation to check if exact answers are reasonable

🎓 How to Estimate

Estimating means finding an answer that is **close** to the exact answer — without doing all the work!

1. **Round** each number to the nearest 10 or 100.
2. **Add or subtract** the rounded numbers.
3. Write your answer with $\approx$ (which means "approximately equal to").

Example: $347 + 285$

Round to the nearest 100: $347 \approx 300$ and $285 \approx 300$.

$300 + 300 = 600$, so $347 + 285 \approx 600$.

Estimation is also a **superpower** for checking your work! If your exact answer is close to the estimate, it's probably correct. If it's far off, go back and check!

> ⓘ TIP
> Rounding to the nearest 10 gives a **closer** estimate, but rounding to the nearest 100 is **faster** — you can do it in your head!

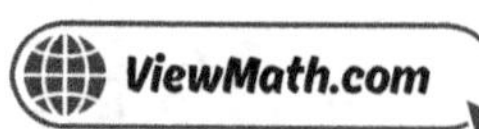

Estimate $463 + 219$ *(Round to Nearest Hundred)*

1. Round: $463 \approx 500$

2. Round: $219 \approx 200$

3. Add: $500 + 200 = 700$

> ✓ **Answer:** $463 + 219 \approx 700$

The exact answer is 682. Our estimate of 700 is close! ✓

Estimate $724 - 389$ *(Round to Nearest Hundred)*

1. Round: $724 \approx 700$

2. Round: $389 \approx 400$

3. Subtract: $700 - 400 = 300$

> ✓ **Answer:** $724 - 389 \approx 300$

The exact answer is 335. Our estimate is in the right ballpark!

✏️ Estimation Practice ✏️

Round to the nearest 100 to estimate.

1. $324 + 478 \approx$ ______________

2. $672 + 281 \approx$ ______________

3. $793 - 348 \approx$ ______________

4. $845 - 412 \approx$ ______________

5. Kim says $352 + 219 = 771$. Estimate to check: is her answer reasonable? __________

Find more at
ViewMath.com/Grade3

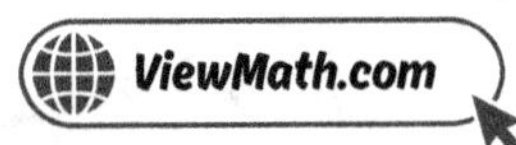

6. *A farmer has 387 sheep and 425 cows. About how many animals does the farmer have? (Round to the nearest hundred.)*

Answer: _________________ *animals*

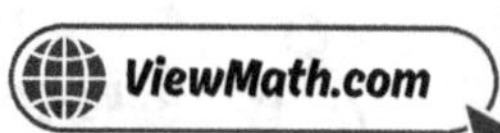

3

Multiplication & Division

What's Inside

★ 3.1 Understanding Multiplication ★

◎ In this lesson you will learn:

- *Understand multiplication as counting equal groups*
- *Read and write multiplication sentences*
- *Use arrays and dot groups to show multiplication*

🎓 What Is Multiplication?

Multiplication is a fast way to add **equal groups**.

3×4 means "3 groups of 4":

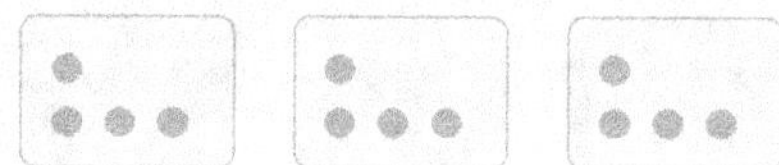

Instead of adding $4 + 4 + 4 = 12$, we simply write $3 \times 4 = 12$.

$$\underbrace{5}_{groups} \times \underbrace{7}_{in\ each} = \underbrace{35}_{total}$$

The numbers we multiply are called **factors**. The answer is called the **product**.

We can also show multiplication as an **array** (rows and columns):

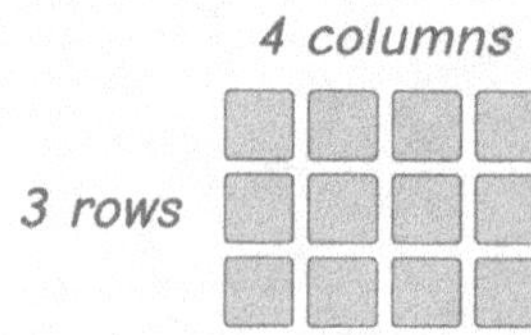

3 rows of $4 = 12$

> 66 *Think about it this way: the first number tells you HOW MANY groups, and the second tells you HOW MANY in each group!* 99

Show 2×6 with Equal Groups

2×6 means "2 groups of 6":

1. We have 2 groups with 6 in each group

2. As repeated addition: $6 + 6 = 12$

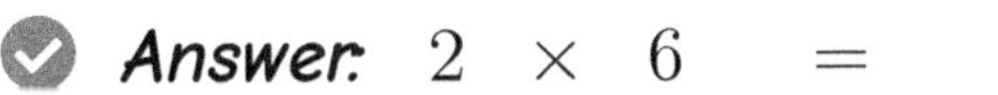

> ✔ **Answer:** $2 \times 6 = 12$

Write a Multiplication Sentence for an Array

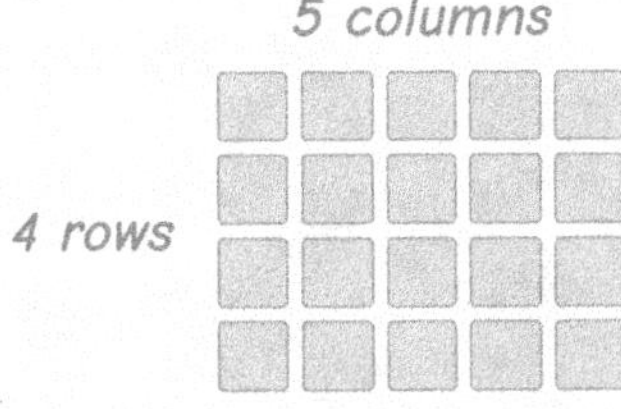

1. Count the rows: 4 rows

2. Count the columns: 5 in each row

3. Write the sentence: 4×5

> ✔ **Answer:** $4 \times 5 - 20$

Multiplication Practice

1. Write the multiplication sentence: 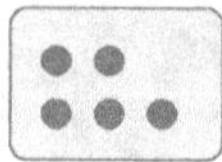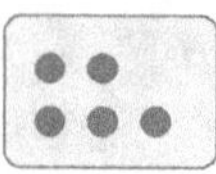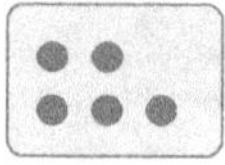 _____________

2. Write the multiplication sentence:  _____________

   7 columns

   2 rows

3. Write as multiplication: $6 + 6 + 6 + 6 =$ _________ $\longrightarrow$ _____________

4. $5 \times 3 =$ _________

5. $4 \times 4 =$ _________

6. There are 5 bags with 4 apples in each bag. How many apples are there in all?

   Answer: _____________ apples

★ 3.2 Multiplication Facts 0–5 ★

◎ In this lesson you will learn:

- Know the special rules for multiplying by 0 and 1
- Use skip counting and doubles to multiply by 2, 3, 4, and 5

🎓 Patterns for Facts 0–5

Multiply by 0: Any number times 0 equals **0**.

$5 \times 0 = 0$ $0 \times 8 = 0$ Zero groups means nothing!

Multiply by 1: Any number times 1 equals **that number**.

$7 \times 1 = 7$ $1 \times 9 = 9$ One group is just the number!

Multiply by 2: Double the number (add it to itself).

$2 \times 6 = 6 + 6 = 12$ $2 \times 8 = 8 + 8 = 16$

Multiply by 5: Skip count by 5s. Products always end in **0** or **5**.

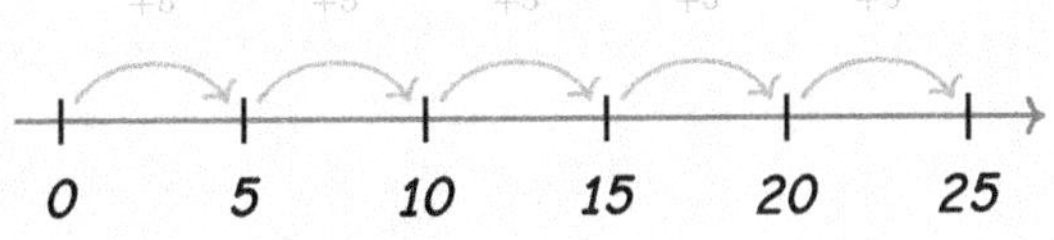

Multiply by 4: Double the double!

4×6: double 6 = 12, double 12 = **24**

Multiply by 3: Skip count by 3s: $3, 6, 9, 12, 15, \ldots$

> *The 5s table is easy to remember — just look at a clock! Each number on the clock is 5 minutes apart: 5, 10, 15, 20, 25, …*

3×4 *as an Array*

3 rows of 4

Count by rows:

Row 1: 4

Row 2: 4

Row 3: 4

Total: $4 + 4 + 4 = 12$

Answer: $3 \times 4 = 12$

Facts 0–5 Practice

1. $0 \times 7 =$ _________

2. $1 \times 9 =$ _________

3. $2 \times 8 =$ _________

4. $5 \times 6 =$ _________

5. $4 \times 7 =$ _________

6. $3 \times 9 =$ _________

7. A classroom has 4 rows of desks. Each row has 5 desks. How many desks are in the classroom?

Answer: _______________ desks

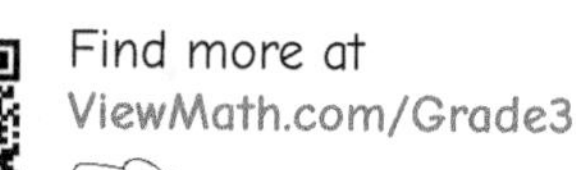

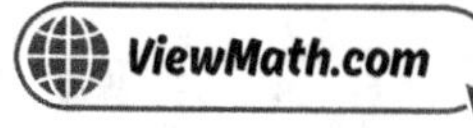

⭐ 3.3 Multiplication Facts 6–10 ⭐

◎ In this lesson you will learn:

- Use strategies and tricks to learn facts for 6, 7, 8, 9, and 10
- Build fluency with the full times table

🎓 Strategies for Facts 6–10

Multiply by 10: Just add a zero to the end!

$6 \times 10 = 60 \qquad 9 \times 10 = 90 \qquad 10 \times 10 - 100$

Multiply by 9 — the Finger Trick:

Hold up all 10 fingers. To find $9 \times n$, put down finger number n.

9×3: put down finger 3. Left side: **2** fingers (tens). Right side: **7** fingers (ones). Answer: **27!**

Bonus: the digits of every 9s product **add up to** 9: $18 \to 1 + 8 = 9, \quad 27 \to 2 + 7 = 9, \quad 36 \to 3 + 6 = 9, \dots$

Multiply by 6: Think $5 \times n + n$ (one more group).

6×7: $5 \times 7 = 35$, plus one more 7: $35 + 7 = 42$.

Multiply by 8: Double, double, double!

8×7: double 7 = 14, double 14 = 28, double 28 = **56**.

Multiply by 7: No special trick, but remember — you already know most 7s facts from earlier tables! The new ones to memorize:

$7 \times 6 = 42 \qquad 7 \times 7 = 49 \qquad 7 \times 8 = 56 \qquad 7 \times 9 = 63$

> 66 After learning facts 0–5 and the tricks for 9 and 10, there are only **6** tough facts left to memorize: $6 \times 6, 6 \times 7, 6 \times 8, 7 \times 7, 7 \times 8,$ and 8×8. You can do it! 99

Find more at
ViewMath.com/Grade3

Use the 9s Finger Trick for 9×4

1. Hold up 10 fingers

2. Put down finger number 4

3. Count fingers on the left: **3** (tens)

4. Count fingers on the right: **6** (ones)

 **Answer:** $9 \times 4 = 36$

Use "One More Group" for 6×8

1. Start with $5 \times 8 = 40$ (a fact you know)

2. Add one more group of 8: $40 + 8 = 48$

 **Answer:** $6 \times 8 = 48$

Facts 6–10 Practice

1. $6 \times 7 =$ _________

2. $9 \times 6 =$ _________

3. $8 \times 8 =$ _________

4. $7 \times 9 =$ _________

5. $10 \times 7 =$ _________

6. $8 \times 6 =$ _________

7. A bookshelf has 7 shelves. Each shelf holds 8 books. How many books fit on the bookshelf?

Answer: _______________ books

 Find more at
ViewMath.com/Grade3

Multiplication Properties Practice

1. $5 \times 9 = 9 \times$ __________

2. If $4 \times 7 = 28$, then $7 \times 4 =$ __________

3. Use the distributive property: $6 \times 8 = (6 \times 5) + (6 \times$ _____$) = 30 +$ _______ $=$ _______

4. $3 \times 8 = 8 \times 3$ True ○   False ○

5. $5 \times 0 = 5$ True ○   False ○

6. Leo knows $5 \times 7 = 35$. How can he use this to find 6×7? What is the answer?

Answer: __________________

★ 3.6 Understanding Division ★

◎ In this lesson you will learn:

- Understand division as sharing equally or making equal groups
- Read and write division sentences

🎓 What Is Division?

Division means splitting a total into **equal groups**. There are two ways to think about it:

Sharing (How many in each group?)

12 cookies shared among 3 friends. How many does each friend get?

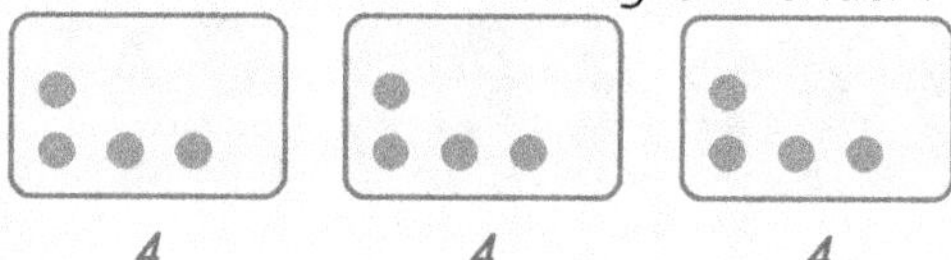

$$12 \div 3 = 4 \quad \text{(each friend gets 4)}$$

Grouping (How many groups?)

12 cookies, put 4 in each bag. How many bags?

$$12 \div 4 = 3 \quad \text{(3 bags)}$$

The three parts of a division sentence:

$$\underbrace{12}_{\text{dividend}} \div \underbrace{3}_{\text{divisor}} = \underbrace{4}_{\text{quotient}}$$

Division is the **opposite** of multiplication! If $3 \times 4 = 12$, then $12 \div 3 = 4$ and $12 \div 4 = 3$.

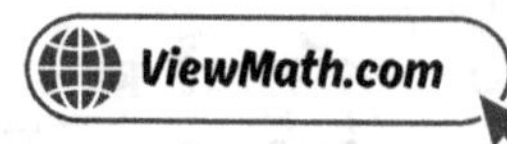

> *Think of a multiplication fact you know, and you get TWO division facts for free!* $5 \times 8 = 40$ *gives you* $40 \div 5 = 8$ *and* $40 \div 8 = 5$.

$20 \div 5 = ?$ *(Sharing)*

Share 20 apples equally among 5 baskets.

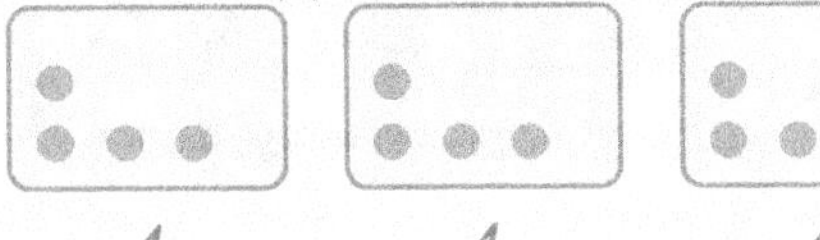
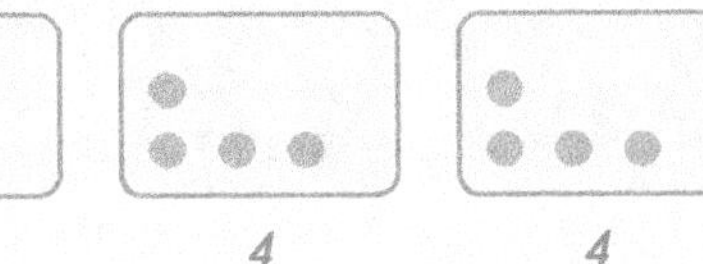

4   4   4   4   4

1. We share 20 items into 5 equal groups

2. Each group gets 4 items

3. Check: $5 \times 4 = 20$ ✓

> ✓ **Answer:** $20 \div 5 = 4$

$56 \div 8 = ?$ *(Think Multiplication!)*

1. Ask: $8 \times ? = 56$

2. We know $8 \times 7 = 56$

3. So $56 \div 8 = 7$

> ✓ **Answer:** $56 \div 8 = 7$

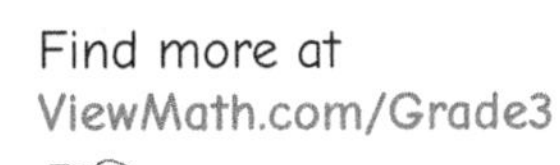

Find more at
ViewMath.com/Grade3

Division Practice

1. $15 \div 3 =$ __________

2. $24 \div 4 =$ __________

3. $42 \div 6 =$ __________

4. $36 \div 9 =$ __________

5. $18 \div 3 = 6$ True ○   False ○

6. There are 32 students. The teacher splits them into 4 equal teams. How many students are on each team?

Answer: __________ students

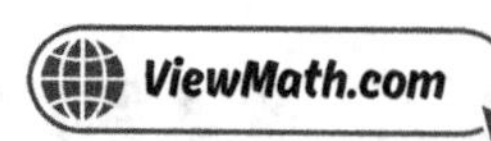

★ 3.7 Division Facts ★

In this lesson you will learn:

- Use *"think multiplication"* to solve division facts quickly
- Know division facts for divisors 1 through 10

Think Multiplication!

The fastest way to divide is to **think multiplication**. Every division fact has a matching multiplication fact:

Division Problem		Think...
$56 \div 8 = ?$	$\longrightarrow$	$8 \times ? = 56$
$56 \div 8 = \mathbf{7}$		$8 \times \mathbf{7} = 56 \checkmark$
$36 \div 4 = ?$	$\longrightarrow$	$4 \times ? = 36$
$36 \div 4 = \mathbf{9}$		$4 \times \mathbf{9} = 36 \checkmark$

If you know your times tables, you already know your division facts!

Each multiplication fact gives you **two** division facts:

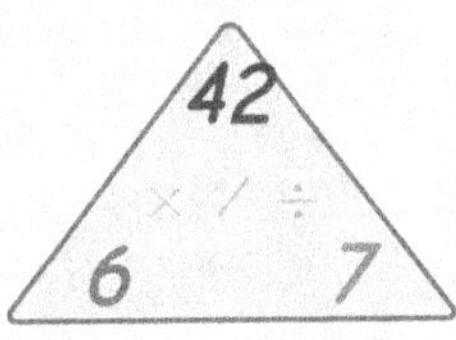

$6 \times 7 = 42 \quad \longrightarrow \quad 42 \div 6 = 7 \quad$ and $\quad 42 \div 7 = 6$

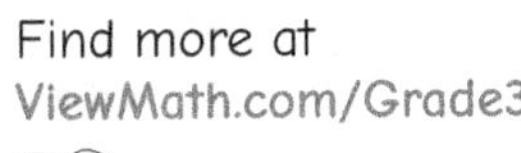
Find more at
ViewMath.com/Grade3

66 *Don't memorize division facts separately — just flip your times tables around! See $48 \div 8$?* *Think: "8 times WHAT equals 48?" Answer: 6!* 99

Using "Think Multiplication"

Solve $63 \div 9 = ?$

1. *Think: "9 times what equals 63?"*

2. *I know $9 \times 7 = 63$*

3. *So $63 \div 9 = 7$*

Answer: $63 \div 9 = 7$

A Tricky Fact

Solve $72 \div 8 = ?$

1. *Think: "8 times what equals 72?"*

2. *I know $8 \times 9 = 72$*

3. *So $72 \div 8 = 9$*

Answer: $72 \div 8 = 9$

Find more at
ViewMath.com/Grade3

Division Facts Practice

1. $45 \div 5 =$ _____________

2. $32 \div 4 =$ _____________

3. $54 \div 6 =$ _____________

4. $49 \div 7 =$ _____________

5. $81 \div 9 =$ _____________

6. A baker made 48 muffins and packed them into boxes of 8. How many boxes did she fill?

Answer: _____________ boxes

⭐ 3.8 Relating Multiplication & Division ⭐

◎ In this lesson you will learn:

- Understand that multiplication and division are **inverse operations**
- Write all four facts in a fact family

🎓 Fact Families

Multiplication and division are **inverse operations** — they undo each other. Three numbers that go together form a **fact family** with 4 related facts:

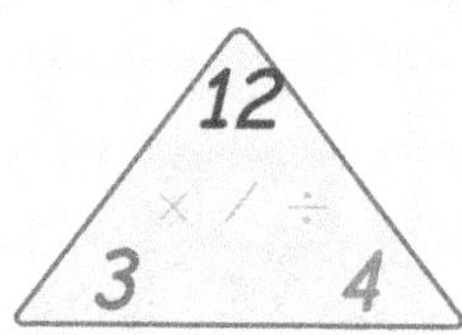

$$3 \times 4 = 12 \qquad\qquad 4 \times 3 = 12$$

$$12 \div 3 = 4 \qquad\qquad 12 \div 4 = 3$$

One multiplication fact gives you **two** division facts! The **largest number** is always the product (in multiplication) and the dividend (in division).

Special Case: Square Numbers

When both factors are the same, the family has only 2 facts:

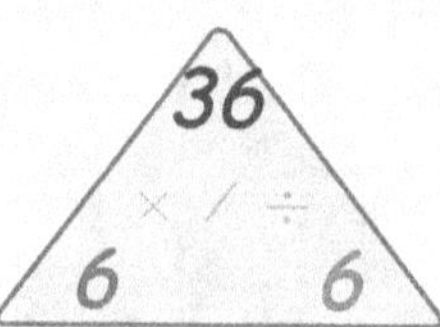

$$6 \times 6 = 36 \qquad 36 \div 6 = 6$$

Find more at
ViewMath.com/Grade3

ViewMath.com

> To check any division answer, just multiply! If $32 \div 8 = 4$, check: $4 \times 8 = 32$ ✓

Writing a Fact Family

Write the fact family for 7, 8, and 56.

1. Identify the largest number: 56 (the product / dividend)

2. Write the multiplication facts: $7 \times 8 = 56$ and $8 \times 7 = 56$

3. Write the division facts: $56 \div 7 = 8$ and $56 \div 8 = 7$

✓ **Answer:** $7 \times 8 = 56, \quad 8 \times 7 = 56, \quad 56 \div 7 = 8, \quad 56 \div 8 = 7$

Relating Multiplication & Division Practice

1. Write all four facts for the fact family $\{5, 9, 45\}$:

____

____

____

____

2. Write the fact family for $\{7, 7, 49\}$ (how many facts?):

____

____

3. $72 \div 8 = 9$.   Check with multiplication: $9 \times 8 =$ __________.   Correct? __________

Find more at
ViewMath.com/Grade3

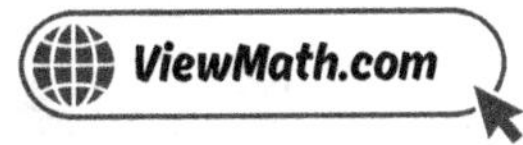

4. $42 \div 7 = 5.$ *Check with multiplication:* $5 \times 7 =$ __________ .   *Correct?* __________

5. $6 \times$ __________ $= 54$

6. $48 \div$ __________ $= 8$

Find more at
ViewMath.com/Grade3

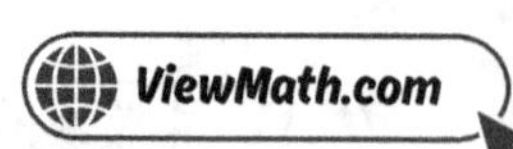

★ 3.9 Find the Missing Number ★

◎ In this lesson you will learn:

- Find unknown numbers in multiplication and division equations
- Use inverse operations to solve for unknowns

🎓 Using Inverse Operations

Any number in a multiplication or division equation can be missing. Use the **inverse operation** to find it!

Missing Factor ⟶ Divide

$$8 \times \mathbf{?} = 48 \qquad \longrightarrow \qquad 48 \div 8 = \mathbf{6}$$

$$\mathbf{?} \times 7 = 63 \qquad \longrightarrow \qquad 63 \div 7 = \mathbf{9}$$

When a **factor** is missing, **divide** the product by the known factor.

Missing Dividend ⟶ Multiply

$$\mathbf{?} \div 3 = 5 \qquad \longrightarrow \qquad 5 \times 3 = \mathbf{15}$$

$$\mathbf{?} \div 9 = 4 \qquad \longrightarrow \qquad 4 \times 9 = \mathbf{36}$$

When the **dividend** is missing, **multiply** the quotient by the divisor.

Missing Divisor ⟶ Divide

$$56 \div \mathbf{?} = 7 \qquad \longrightarrow \qquad 56 \div 7 = \mathbf{8}$$

Find more at
ViewMath.com/Grade3

> **TIP**
> Always **check** your answer by plugging it back in!
>
> $8 \times ? = 48 \quad \rightarrow \quad ? = 6 \quad \rightarrow \quad 8 \times 6 = 48 \checkmark$

Missing Factor

Solve $n \times 6 = 42$.

1. The product is 42 and one factor is 6

2. Divide: $42 \div 6 = 7$

3. Check: $7 \times 6 = 42 \checkmark$

✔ **Answer:** $n \quad = \quad 7$

Missing Dividend

Solve $n \div 4 = 7$.

1. The divisor is 4 and the quotient is 7

2. Multiply: $7 \times 4 = 28$

3. Check: $28 \div 4 = 7 \checkmark$

✔ **Answer:** $n \quad = \quad 28$

Find more at
ViewMath.com/Grade3

Find the Missing Number Practice

1. $n \times 5 = 35$ $\quad n =$ _________

2. $8 \times n = 72$ $\quad n =$ _________

3. $n \div 6 = 9$ $\quad n =$ _________

4. $48 \div n = 8$ $\quad n =$ _________

5. $n \times 9 = 36$ $\quad n =$ _________

6. Sara has some stickers. She gives 7 stickers to each of 6 friends and has none left. How many stickers did she start with? Write an equation and solve.

Answer: _________________ stickers

Find more at
ViewMath.com/Grade3

3.10 Multiplication & Division Word Problems

In this lesson you will learn:

- Identify whether a word problem needs multiplication or division
- Use clue words to choose the correct operation

Multiply or Divide?

Read the problem carefully and look for **clue words** to decide which operation to use.

Multiply ×	Divide ÷
You know the **groups** and the **size** of each group. Find the **total**.	You know the **total**. Find the group size or number of groups.
Clue words: each, per, every, times as many, rows of	**Clue words:** share equally, split, how many in each, how many groups
"4 bags with 6 apples each" → $4 \times 6 = 24$	"24 apples shared among 4 bags" → $24 \div 4 = 6$

66 Read the problem twice! First to understand the story, then to pick the right operation. 99

Find more at
ViewMath.com/Grade3

Multiplication Word Problem

There are 7 tables in the lunchroom. Each table has 4 chairs. How many chairs are there in all?

1. We know the groups (7 tables) and the size (4 chairs each)

2. Clue word "each" tells us to **multiply**

3. $7 \times 4 = 28$

 **Answer:** 28 **chairs**

Division Word Problem

Mom baked 36 cookies. She shares them equally among 4 friends. How many cookies does each friend get?

1. We know the total (36) and the number of groups (4 friends)

2. Clue words "shares equally" tell us to **divide**

3. $36 \div 4 = 9$

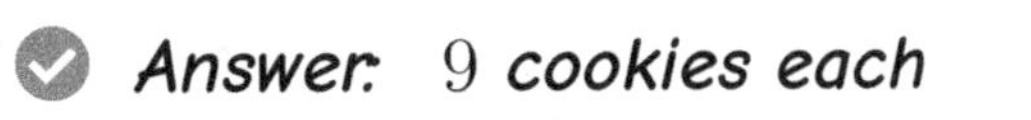 **Answer:** 9 **cookies each**

✏ Word Problems Practice ✏

1. A bookshelf has 5 shelves. Each shelf holds 9 books. How many books are on the bookshelf?

   Answer: _________________ books

2. Sam rides his bike 6 miles every day. How many miles does he ride in 7 days?

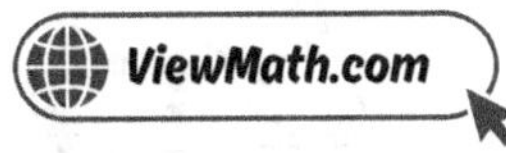

Answer: _______________ miles

3. There are 42 students. The teacher splits them into 6 equal teams. How many students are on each team?

Answer: _______________ students

4. Maria has 35 beads. She uses 5 beads per bracelet. How many bracelets can she make?

Answer: _______________ bracelets

5. Ana has 8 crayons. Ben has 3 times as many crayons as Ana. How many crayons does Ben have?

Answer: _______________ crayons

6. A baker has 54 muffins. He packs 9 muffins in each box. How many boxes does he fill?

Answer: _______________ boxes

★ 3.11 Two-Step Word Problems ★

◎ In this lesson you will learn:

- Solve word problems that need two steps
- Use a letter to represent the unknown in an equation

🎓 What Is a Two-Step Problem?

A **two-step problem** needs **two operations** to solve. First answer a **hidden question**, then use that answer to find the final answer.

Example: Emma buys 3 packs of pencils with 8 pencils each. She gives 5 pencils to a friend. How many pencils does Emma have left?

Step 1 (hidden question): How many pencils total?

$3 \times 8 = 24$

Step 2 (final answer): Subtract the 5 she gave away.

$24 - 5 = 19$

You can write this as **one equation** with a letter:

$$p = (3 \times 8) - 5 = 24 - 5 = 19$$

The **parentheses** show which part to do first!

❝ Ask yourself: "What do I need to figure out BEFORE I can answer the main question?" That's your hidden question! ❞

Find more at
ViewMath.com/Grade3

Multiply Then Add

A farmer has 4 rows of apple trees with 7 trees in each row. He plants 6 more trees. How many trees does he have now?

1. Hidden question: How many trees in the rows? $4 \times 7 = 28$

2. Final answer: Add the extra trees. $28 + 6 = 34$

3. As one equation: $n = (4 \times 7) + 6 = 34$

Answer: **34 trees**

Add Then Divide

Jack picks 15 oranges in the morning and 9 in the afternoon. He puts them equally into 6 bags. How many oranges are in each bag?

1. Hidden question: How many oranges total? $15 + 9 = 24$

2. Final answer: Divide into bags. $24 \div 6 = 4$

3. As one equation: $n = (15 + 9) \div 6 = 4$

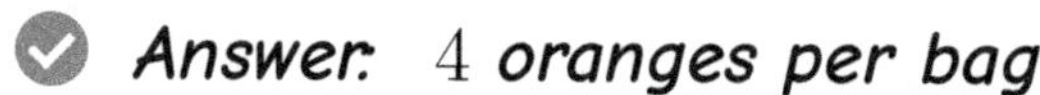

Answer: **4 oranges per bag**

✏ Two-Step Word Problems Practice ✏

1. Sara buys 4 boxes of crayons with 8 crayons in each box. She already has 12 crayons at home. How many crayons does she have in all?

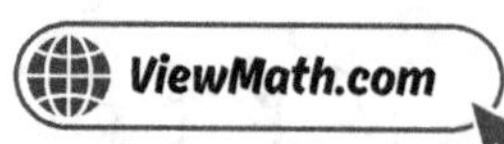

Answer: _________________ crayons

2. A farmer collects 7 baskets of eggs with 6 eggs in each basket. He gives 10 eggs to his neighbor. How many eggs does he have left?

Answer: _________________ eggs

3. There are 24 students split into 4 equal groups. Then 3 more students join one group. How many students are in that group now?

Answer: _________________ students

4. Jake has 40 marbles shared equally among 5 friends. Each friend then loses 2 marbles. How many marbles does each friend have now?

Answer: _________________ marbles

5. Emma picks 16 flowers and her brother picks 8 flowers. They put all the flowers equally into 3 vases. How many flowers are in each vase?

Answer: _________________ flowers

6. A store had 50 toy cars. They sold 14 cars. They put the rest equally on 4 shelves. How many cars are on each shelf?

Answer: _________________ cars

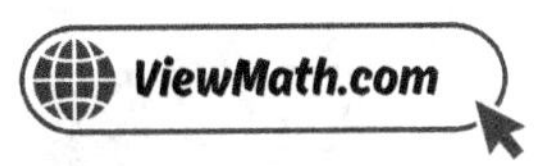

3.12 Patterns in Tables

In this lesson you will learn:

- Find the rule in input/output tables
- Identify patterns in multiplication tables

Input/Output Tables and Patterns

An **input/output table** follows a rule. Find the rule by looking at how the input changes to become the output.

In	Out
2	6
4	12
5	15
7	21

Each output is the input $\times$ 3. The rule is **multiply by** 3.

Patterns in the multiplication table:

- **Multiples of** 2 are always **even**.
- **Multiples of** 5 end in 0 or 5.
- **Multiples of** 9: the digits always add up to 9 (e.g. $18 \rightarrow 1 + 8 = 9$).
- $3 \times 7 = 7 \times 3$ — the table is the same on both sides of the diagonal (**commutative property**).

66 To find the rule, ask: "What do I do to every input to get the output?" Try adding, subtracting, or multiplying! 99

Find more at
ViewMath.com/Grade3

Find the Rule

What is the rule? Find the missing output.

In	Out
3	15
5	25
6	30
8	?

1. Check: $3 \times ? = 15 \quad \rightarrow \quad 3 \times 5 = 15$ ✓

2. Check: $5 \times 5 = 25$ ✓ and $6 \times 5 = 30$ ✓

3. The rule is **multiply by** 5

4. Missing output: $8 \times 5 = 40$

 **Answer: Rule:** × 5. **Missing output:** 40

Patterns in Tables Practice

1. *Find the rule and fill in the missing number.*

In	Out
2	8
4	16
6	24
9	___

Rule: _______________

Find more at
ViewMath.com/Grade3

2. *Find the rule and fill in the missing number.*

In	Out
3	21
5	35
8	56
10	_______

Rule: _________________

3. *What are the next three numbers in this pattern?*

6, 12, 18, 24, ______, ______, ______   *Rule:* _________________

4. *What are the next three numbers in this pattern?*

9, 18, 27, 36, ______, ______, ______   *Rule:* _________________

5. *Fill in the missing products in this piece of the multiplication table.*

×	3	5	7	9
4	_____	20	_____	_____
6	18	_____	42	_____

6. *Circle the numbers that are multiples of 9. (Hint: add the digits — if they sum to 9, it's a multiple of 9!)*

18   25   36   44   54   63   70   81

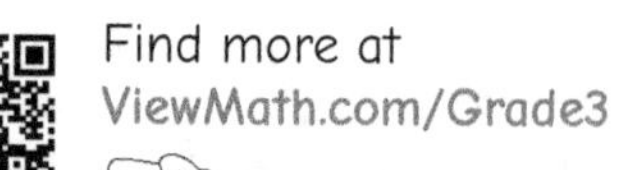

⭐ 3.13 Multiplying 2-Digit by 1-Digit ⭐

 **In this lesson you will learn:**

- Multiply a 2-digit number by a 1-digit number
- Use the **break-apart (distributive)** strategy to find the product

🎓 The Break-Apart Strategy

To multiply a 2-**digit number** by a 1-**digit number**, break the 2-digit number into **tens** and **ones**. Multiply each part, then add the **partial products**.

Example: 4×23

1. Break 23 into $20 + 3$.
2. Multiply each part by 4:

$$4 \times 20 = 80 \qquad \text{and} \qquad 4 \times 3 = 12$$

3. Add the partial products:

$$80 + 12 = 92$$

So $4 \times 23 = 92$.

This works because of the **distributive property**:

$$4 \times 23 = (4 \times 20) + (4 \times 3)$$

> ❝ Break the big number into friendly parts! Tens are easy to multiply, and ones are easy too. Then just add! ❞

3×45

1. Break 45 into $40 + 5$.

2. Multiply each part: $3 \times 40 = 120$ and $3 \times 5 = 15$.

3. Add: $120 + 15 = 135$.

Answer: $3 \quad \times \quad 45 \quad = \quad 135$

6×17

1. Break 17 into $10 + 7$.

2. Multiply each part: $6 \times 10 = 60$ and $6 \times 7 = 42$.

3. Add: $60 + 42 = 102$.

Answer: $6 \quad \times \quad 17 \quad = \quad 102$

2-Digit × 1-Digit Practice

1. $2 \times 34 = $ _________

2. $5 \times 18 = $ _________

3. $4 \times 26 = $ _________

4. $7 \times 13 = $ _________

5. $3 \times 52 = $ _________

6. $8 \times 15 = $ _________

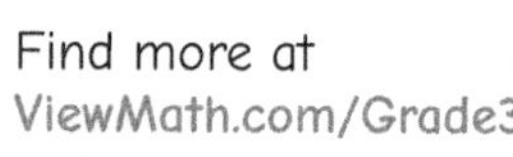
Find more at
ViewMath.com/Grade3

4

Fractions

What's Inside

★ 4.1 Understanding Fractions ★

◎ In this lesson you will learn:

- Understand that fractions come from splitting a whole into equal parts
- Know what the numerator and denominator mean

🎓 What Is a Fraction?

A **fraction** tells you how many **equal parts** you have out of a whole.

Look at this bar. It is split into 4 equal parts. 3 parts are shaded:

We write this as $\frac{3}{4}$ and say "three-fourths."

Every fraction has two numbers:

$$\overbrace{3}^{\text{numerator}}$$
$$\underbrace{4}_{\text{denominator}}$$

- The **numerator** (top) tells how many parts are shaded.
- The **denominator** (bottom) tells how many equal parts the whole is split into.

Here is a circle split into 4 equal parts with 1 part shaded:

This is $\frac{1}{4}$ — one-fourth.

Important: The parts must be **equal**! If the parts are different sizes, you cannot write a fraction.

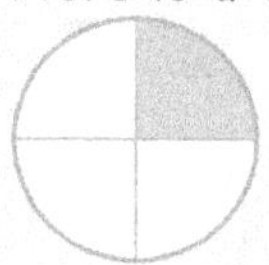

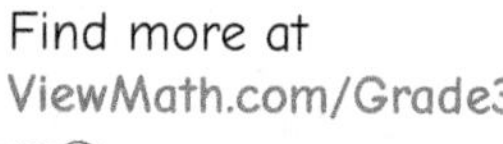

Find more at
ViewMath.com/Grade3

> *Remember: the Denominator is Down on the bottom. It tells how many parts the whole is Divided into!*

What Fraction Is Shaded?

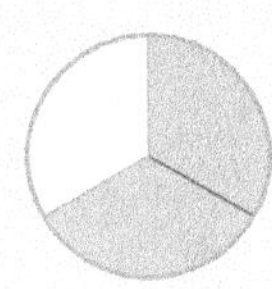

1. Count the total equal parts: 3. That is the denominator.

2. Count the shaded parts: 2. That is the numerator.

3. Write the fraction: $\frac{2}{3}$

> ✅ **Answer:** $\frac{2}{3}$ — two-thirds

✏️ Fraction Practice ✏️

1. The bar below shows a fraction. Write the fraction.

   __________

2. The circle below shows a fraction. Write the fraction.

____

3. In the fraction $\frac{5}{8}$, what is the numerator? __________

4. In the fraction $\frac{5}{8}$, what is the denominator? __________

5. A rectangle is split into 6 equal parts. 5 parts are shaded. Write the fraction. __________

6. A pizza is cut into 8 equal slices. Emma eats 3 slices. What fraction of the pizza did Emma eat?

Answer: _________________ *of the pizza*

⭐ 4.2 Fractions on a Number Line ⭐

◎ In this lesson you will learn:

- Place fractions on a number line from 0 to 1
- Use the denominator to divide the line into equal parts

🎓 Fractions on a Number Line

Fractions are numbers that live **between** whole numbers on a number line.

To place a fraction on a number line:

1. Look at the **denominator** — divide the space from 0 to 1 into that many equal parts.
2. Look at the **numerator** — count that many parts from 0.
3. Mark the spot!

Here is a number line divided into **4** equal parts (fourths):

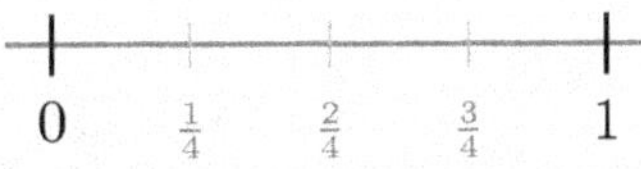

Each mark is $\frac{1}{4}$ of the way from 0 to 1.
The fraction $\frac{3}{4}$ is at the **third** mark from 0.

Compare this to a fraction bar:

Both visuals show the same fraction — $\frac{3}{4}$ means 3 out of 4 equal parts!

> 66 The denominator is like a ruler — it tells you how finely to divide the number line. More parts means smaller pieces! 99

Find more at
ViewMath.com/Grade3

Place $\frac{2}{6}$ on a Number Line

1. The denominator is 6, so divide 0 to 1 into 6 equal parts.

2. The numerator is 2, so count 2 parts from 0.

3. Mark the second tick mark — that is $\frac{2}{6}$!

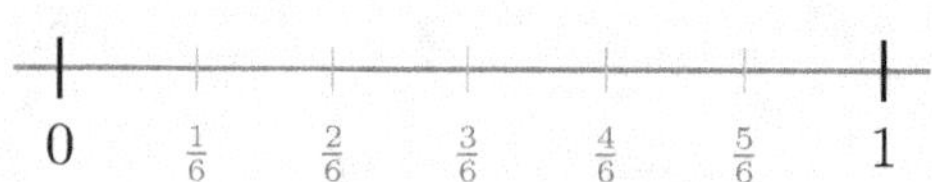

> ✔ **Answer:** $\frac{2}{6}$ **is at the second mark from** 0.

✏ Number Line Practice ✏

1. This number line is divided into 3 equal parts. What fraction is at the first mark?

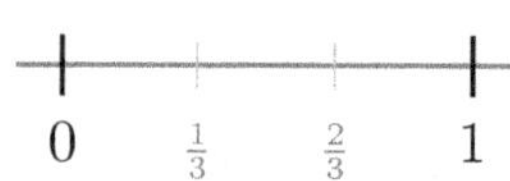

____

2. This number line is divided into 4 equal parts. What fraction is at the second mark?

____

3. How many equal parts do you need to show **sixths** on a number line? __________

4. Where is $\frac{4}{4}$ on the number line? __________

5. $\frac{0}{8}$ is the same as what whole number? __________

6. Maya walks from home to school. The path is divided into 4 equal parts. She has walked 3 parts. What fraction of the path has she walked?

Find more at
ViewMath.com/Grade3

ViewMath.com

Answer: _______________ *of the path*

⭐ 4.3 Unit Fractions ⭐

◉ In this lesson you will learn:

- *Know that a unit fraction has 1 as the numerator*
- *Understand that a bigger denominator means a smaller piece*

🎓 What Is a Unit Fraction?

A **unit fraction** is a fraction with **1** on top. It represents **one single part** of a whole.

Fraction	Visual	Meaning
$\frac{1}{2}$		1 part out of 2
$\frac{1}{4}$		1 part out of 4
$\frac{1}{8}$		1 part out of 8

Key idea: The **bigger** the denominator, the **smaller** each piece!

Compare $\frac{1}{2}$ and $\frac{1}{8}$: when you split a whole into 2 parts, each part is big. When you split the same whole into 8 parts, each part is tiny!

$$\frac{1}{2} > \frac{1}{3} > \frac{1}{4} > \frac{1}{6} > \frac{1}{8}$$

Think of sharing a pizza: 2 people sharing means big slices ($\frac{1}{2}$ each). 8 people sharing means tiny slices ($\frac{1}{8}$ each). More people = more parts = smaller pieces!

> *Unit fractions are the "building blocks" of all fractions! For example, $\frac{3}{4}$ is just three copies of $\frac{1}{4}$.*

Which Is Larger: $\frac{1}{3}$ or $\frac{1}{6}$?

1. Draw both fractions:

   $\frac{1}{3}$:

   $\frac{1}{6}$:

2. The bar for $\frac{1}{3}$ has a bigger shaded piece than $\frac{1}{6}$.

3. Fewer parts means each part is larger!

✓ **Answer:** $\frac{1}{3}$ $>$ $\frac{1}{6}$

✏ Unit Fraction Practice ✏

1. Is $\frac{1}{5}$ a unit fraction? _____________

2. Is $\frac{3}{4}$ a unit fraction? _____________

3. Circle the larger fraction: $\frac{1}{2}$ or $\frac{1}{4}$ _____________

4. Circle the larger fraction: $\frac{1}{3}$ or $\frac{1}{8}$ _____________

5. Order from largest to smallest: $\frac{1}{4}$, $\frac{1}{2}$, $\frac{1}{8}$ _____________

6. Two friends share a ribbon equally. Each friend gets $\frac{1}{2}$ of the ribbon. If 4 friends share the same ribbon equally, each friend gets $\frac{1}{4}$. Which piece is longer — $\frac{1}{2}$ or $\frac{1}{4}$?

   Answer: _____________ of the ribbon

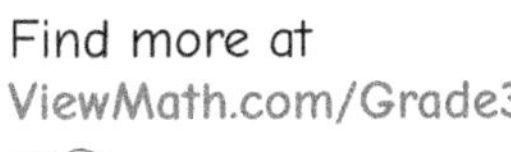

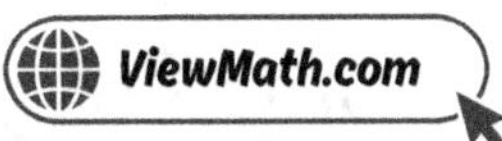

⭐ 4.4 Building Fractions ⭐

◎ In this lesson you will learn:

- Build fractions by counting unit fractions
- Understand that $\frac{a}{b}$ means a copies of $\frac{1}{b}$

🎓 Building Fractions from Unit Fractions

*Every fraction is made by **counting** unit fractions!*

$\frac{3}{4}$ *means 3 copies of* $\frac{1}{4}$*:*

$$\underbrace{\frac{1}{4} + \frac{1}{4} + \frac{1}{4}}_{3 \text{ copies}} = \frac{3}{4}$$

*On a number line, building a fraction means **hopping** from 0. Each hop is one unit fraction long.*
To show $\frac{3}{4}$*, take 3 hops of* $\frac{1}{4}$*:*

$$\begin{array}{ccccc} 0 & \frac{1}{4} & \frac{2}{4} & \frac{3}{4} & 1 \end{array}$$

Hop 1 → $\frac{1}{4}$ Hop 2 → $\frac{2}{4}$ Hop 3 → $\frac{3}{4}$ ✓

More examples:
- $\frac{2}{3} = \frac{1}{3} + \frac{1}{3}$ (2 copies of $\frac{1}{3}$)
- $\frac{5}{6} = \frac{1}{6} + \frac{1}{6} + \frac{1}{6} + \frac{1}{6} + \frac{1}{6}$ (5 copies of $\frac{1}{6}$)
- $\frac{4}{4} = \frac{1}{4} + \frac{1}{4} + \frac{1}{4} + \frac{1}{4} = 1$ *whole!*

> Building fractions is like stacking LEGO bricks! Each unit fraction is one brick. Stack 3 bricks of $\frac{1}{4}$ and you get $\frac{3}{4}$!

Build $\frac{5}{8}$ from Unit Fractions

$\frac{5}{8}$ means 5 copies of $\frac{1}{8}$.

1. Start with $\frac{1}{8}$ — shade 1 part.

2. Add another $\frac{1}{8}$ — now $\frac{2}{8}$.

3. Keep going: $\frac{3}{8}$, $\frac{4}{8}$, $\frac{5}{8}$.

4. We used 5 copies of $\frac{1}{8}$.

✔ **Answer:** $\frac{5}{8} = \frac{1}{8} + \frac{1}{8} + \frac{1}{8} + \frac{1}{8} + \frac{1}{8}$

✏ Building Fractions Practice ✏

1. $\frac{1}{3} + \frac{1}{3} =$ _________

2. $\frac{1}{4} + \frac{1}{4} + \frac{1}{4} =$ _________

3. $\frac{1}{6} + \frac{1}{6} + \frac{1}{6} + \frac{1}{6} =$ _________

4. Write $\frac{3}{8}$ as a sum of unit fractions: _________________

5. How many copies of $\frac{1}{4}$ make $\frac{4}{4}$? _________

6. Carlos drinks $\frac{1}{3}$ of a water bottle in the morning and another $\frac{1}{3}$ in the afternoon. What fraction of the bottle has he drunk in total?

Answer: _____________ of the bottle

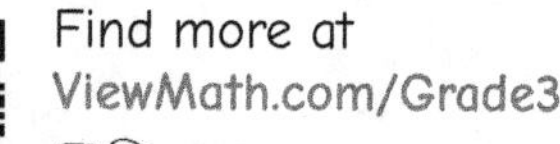
Find more at
ViewMath.com/Grade3

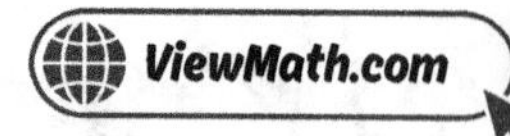

⭐ 4.5 Equivalent Fractions ⭐

◎ In this lesson you will learn:

- Understand that different fractions can name the same amount
- Find equivalent fractions by multiplying the numerator and denominator by the same number

🎓 Same Amount, Different Names

Equivalent fractions are fractions that look different but show the **same amount**.

Look at these fraction bars — the shaded parts are the same size!

$$\frac{1}{2} = \frac{2}{4} = \frac{3}{6}$$

How to Find Equivalent Fractions

Multiply the **numerator** and **denominator** by the **same number**:

$$\frac{1}{2} = \frac{1\times2}{2\times2} = \frac{2}{4} \qquad \frac{1}{2} = \frac{1\times3}{2\times3} = \frac{3}{6} \qquad \frac{1}{2} = \frac{1\times4}{2\times4} = \frac{4}{8}$$

The fraction looks different, but the amount stays the **same**!

Find more at
ViewMath.com/Grade3

> *Equivalent fractions are like nicknames! "William" and "Will" are different names for the same person. $\frac{1}{2}$ and $\frac{2}{4}$ are different names for the same amount!*

Find a Fraction Equivalent to $\frac{1}{3}$

Multiply the numerator and denominator by 2:

1. $\frac{1}{3} = \frac{1 \times 2}{3 \times 2} = \frac{2}{6}$

2. Check with fraction bars:

$\frac{1}{3}$:

$\frac{2}{6}$:

Same shading — they are equivalent! ✓

✔ Answer: $\frac{1}{3} \quad = \quad \frac{2}{6}$

Are $\frac{2}{3}$ and $\frac{4}{6}$ Equivalent?

1. Start with $\frac{2}{3}$. Multiply top and bottom by 2:

2. $\frac{2 \times 2}{3 \times 2} = \frac{4}{6}$ ✓

3. Yes! $\frac{2}{3} = \frac{4}{6}$

✔ Answer: Yes — $\frac{2}{3} \quad = \quad \frac{4}{6}$

✏️ Equivalent Fractions Practice ✏️

1. Find an equivalent fraction: $\frac{1}{4} = \frac{?}{8}$? = ___________

2. Find an equivalent fraction: $\frac{2}{3} = \frac{?}{6}$? = ___________

3. Find an equivalent fraction: $\frac{1}{2} = \frac{?}{8}$? = ___________

4. Are $\frac{3}{4}$ and $\frac{6}{8}$ equivalent? Write Yes or No. ___________

5. Find an equivalent fraction: $\frac{3}{4} = \frac{?}{8}$? = ___________

6. Emma eats $\frac{2}{4}$ of a granola bar. Leo eats $\frac{1}{2}$ of the same size granola bar. Did they eat the same amount? Explain using equivalent fractions.

Answer: _______________ (Yes or No)

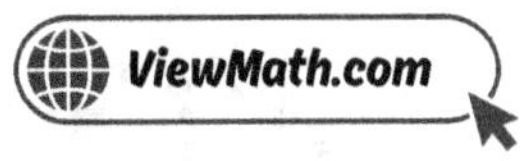

★ 4.6 Whole Numbers as Fractions ★

◎ In this lesson you will learn:

- Write whole numbers as fractions ($3 = \frac{3}{1}$)
- Understand that $\frac{n}{n} = 1$ and $\frac{0}{n} = 0$

🎓 Three Key Rules

Every whole number can be written as a fraction!

Rule 1: Any Whole Number Over 1

Any whole number n equals $\frac{n}{1}$:

$$1 = \tfrac{1}{1} \qquad 2 = \tfrac{2}{1} \qquad 3 = \tfrac{3}{1} \qquad 5 = \tfrac{5}{1}$$

Why? $\frac{3}{1}$ means "3 parts where each part is 1 whole." That's just 3!

Rule 2: Same Top and Bottom = 1 Whole

When the **numerator equals the denominator**, the fraction equals **1**:

$$\frac{2}{2} \quad = \quad 1$$

$$\frac{4}{4} \quad = \quad 1$$

$$\frac{6}{6} \quad = \quad 1$$

All the parts are shaded — you have the **whole thing!**

Rule 3: Zero on Top = 0

$\frac{0}{n} = 0$ for any number n: $\qquad \frac{0}{3} = 0 \qquad \frac{0}{8} = 0$

Zero parts means **nothing!**

Find more at
ViewMath.com/Grade3

> 66 *Remember:* $\frac{5}{5} = 1$, *NOT 5! Same top and bottom always equals 1 whole.* 99

Write 3 as a Fraction with Denominator 4

Each whole has 4 fourths. So 3 wholes have:

1. 1 whole $= \frac{4}{4}$

2. 3 wholes $= 3 \times 4 = 12$ fourths

3. $3 = \frac{12}{4}$

> ✓ **Answer:** $3 \quad = \quad \frac{12}{4}$

What Whole Number Does $\frac{6}{3}$ Equal?

1. $\frac{6}{3}$ means 6 thirds.

2. Each whole has 3 thirds, so $6 \div 3 = 2$ wholes.

> ✓ **Answer:** $\frac{6}{3} \quad = \quad 2$

✏ Whole Numbers as Fractions Practice ✏

1. Write as a fraction: $4 = \frac{?}{1}$ $? =$ ___________

2. What whole number does $\frac{8}{8}$ equal? ___________

3. $\frac{0}{6} =$ ___________

Find more at
ViewMath.com/Grade3

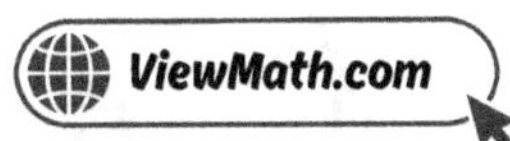

4. Write 2 as a fraction with denominator 3: $2 = \frac{?}{3}$ $? =$ __________

5. What whole number does $\frac{10}{2}$ equal? __________

6. A pie is cut into 6 equal slices. Maria eats all 6 slices. What fraction of the pie did she eat? How many whole pies is that?

Answer: __________________ whole pie(s)

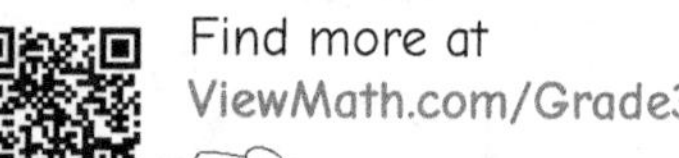

Find more at
ViewMath.com/Grade3

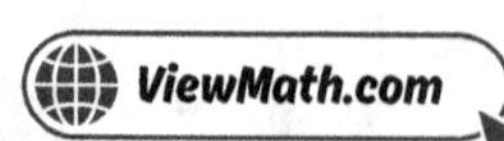

★ 4.7 Comparing Fractions ★

◎ In this lesson you will learn:

- Compare fractions with the same denominator
- Compare fractions with the same numerator
- Use $<$, $>$, and $=$ to write comparisons

🎓 Two Strategies for Comparing Fractions

Strategy 1: Same Denominator — Compare the Numerators

When the pieces are the **same size**, just count how many you have!

$\frac{3}{8}$ vs. $\frac{5}{8}$:

5 pieces > 3 pieces, so $\frac{5}{8} > \frac{3}{8}$

Bigger numerator = bigger fraction (same denominator).

Strategy 2: Same Numerator — Compare the Denominators

When you have the **same number** of pieces, compare piece **sizes**!

$\frac{1}{4}$ vs. $\frac{1}{8}$:

Fourths are **bigger** pieces than eighths, so $\frac{1}{4} > \frac{1}{8}$

Bigger denominator = smaller pieces = smaller fraction (same numerator).

66 *Think of pizza! If you cut a pizza into 4 slices, each slice is WAY bigger than if you cut it into 8 slices. So $\frac{1}{4} > \frac{1}{8}$!* 99

Compare $\frac{2}{6}$ and $\frac{4}{6}$

1. Same denominator? Yes — both are sixths.

2. Compare numerators: 2 vs. 4.

3. $2 < 4$, so $\frac{2}{6} < \frac{4}{6}$.

Answer: $\frac{2}{6}$ $<$ $\frac{4}{6}$

Compare $\frac{3}{4}$ and $\frac{3}{8}$

1. Same numerator? Yes — both have 3 parts.

2. Compare denominators: 4 vs. 8.

3. Fourths are bigger pieces than eighths.

4. 3 big pieces > 3 small pieces.

Answer: $\frac{3}{4}$ $>$ $\frac{3}{8}$

Find more at
ViewMath.com/Grade3

✏️ Comparing Fractions Practice ✏️

1. Compare. Write $<$, $>$, or $=$: $\frac{2}{8}$ _______ $\frac{5}{8}$

2. Compare. Write $<$, $>$, or $=$: $\frac{3}{4}$ _______ $\frac{1}{4}$

3. Compare. Write $<$, $>$, or $=$: $\frac{1}{3}$ _______ $\frac{1}{6}$

4. Compare. Write $<$, $>$, or $=$: $\frac{2}{6}$ _______ $\frac{2}{3}$

5. Compare. Write $<$, $>$, or $=$: $\frac{4}{4}$ _______ $\frac{4}{8}$

6. Jake ate $\frac{3}{6}$ of his sandwich. Amy ate $\frac{5}{6}$ of her same-size sandwich. Who ate more?

Answer: _________________ (Jake or Amy)

Find more at
ViewMath.com/Grade3

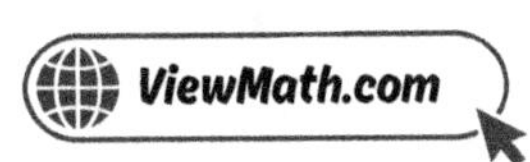

CHAPTER

5

Time, Measurement & Money

⭐ What's Inside ⭐

★ 5.1 Telling Time ★

◎ In this lesson you will learn:

- Read an analog clock to tell time to the nearest minute
- Understand the hour hand and minute hand

🎓 Reading an Analog Clock

An **analog clock** has a round face with numbers 1 through 12 and two hands:

- The **short hand** (hour hand) points to the **hour**.
- The **long hand** (minute hand) points to the **minutes**.

Each number on the clock stands for 5 minutes. **Skip count by** 5s starting from 12:

12	1	2	3	4	5
0	5	10	15	20	25
6	7	8	9	10	11
30	35	40	45	50	55

The small tick marks between numbers are 1 minute each. Count by 5s to the nearest number, then count by 1s for the extra ticks.

> 66 The **short** hand is for **hours** — both are short words! The **long** hand is for **minutes** — you need more minutes to fill an hour! 99

What Time Is It?

1. The short hand is past the 7 → the hour is 7.

2. The long hand is on the 3. Count by 5s: $5, 10, 15$ → 15 minutes.

✓ **Answer:** **The time is** $7{:}15.$

Reading Minutes Between Numbers

1. Short hand is past the 9 → hour is 9.

2. Long hand is between 7 and 8. Count by 5s to the 7: $5, 10, 15, 20, 25, 30, 35.$

3. Then count 2 more tick marks: $36, 37.$

✓ **Answer:** **The time is** $9{:}37.$

Find more at
ViewMath.com/Grade3

✏ Telling Time Practice ✏

1. The short hand is on 4 and the long hand is on 12. What time is it? ___________

2. The short hand is past 6 and the long hand is on the 6. What time is it? ___________

3. The short hand is past 10 and the long hand is on the 3. What time is it? ___________

4. The short hand is past 2 and the long hand is on the 9. What time is it? ___________

5. The short hand is past 8 and the long hand is 2 tick marks past the 5. What time is it? ___________

6. Sam looks at the clock when he starts his homework. The short hand is past the 3 and the long hand is on the 10. What time did Sam start his homework?

Answer: _______________ time

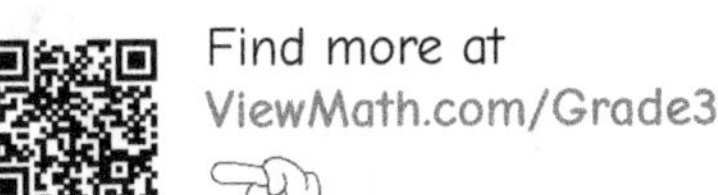

5.2 Elapsed Time ★

◎ In this lesson you will learn:

- *Find how much time has passed between two times*
- *Count forward in hours and minutes to find elapsed time*

🎓 What Is Elapsed Time?

Elapsed time is the amount of time that passes from a **start time** to an **end time**.

To find elapsed time, **count forward** from the start time:

1. First, count by **hours** until you are close.
2. Then, count by **minutes** to reach the end time.
3. Add up: hours + minutes = elapsed time.

Example: From 2:15 to 4:45:

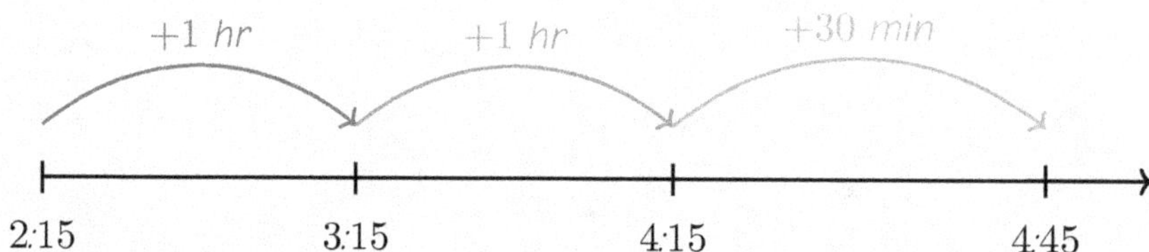

1 hr +1 hr +30 min = **2 hours and 30 minutes**.

❝ When the start and end time are in the **same hour**, just subtract the minutes! For example, 10:45 − 10:15 = 30 minutes. ❞

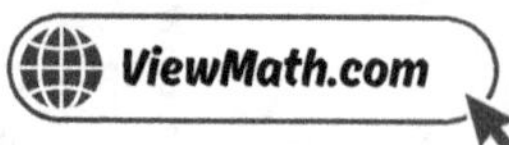

How Long Was the Movie?

A movie started at 4:10 and ended at 6:35. How long was the movie?

1. 4:10 → 5:10 = 1 hour.

2. 5:10 → 6:10 = 1 more hour.

3. 6:10 → 6:35 = 25 minutes.

4. Total: $1 + 1 = 2$ hours and 25 minutes.

 **Answer:** 2 **hours and** 25 **minutes**

Finding the End Time

Soccer practice starts at 3:45 and lasts 1 hour and 30 minutes. When does it end?

1. Start: 3:45.

2. Add 1 hour: 3:45 → 4:45.

3. Add 30 minutes: 4:45 → 5:15.

 **Answer:** 5:15

Elapsed Time Practice

1. How much time passes from 1:00 to 3:30? _________________

2. How much time passes from 8:15 to 8:50? _________________

3. How much time passes from 11:20 to 2:20? _________________

Find more at
ViewMath.com/Grade3

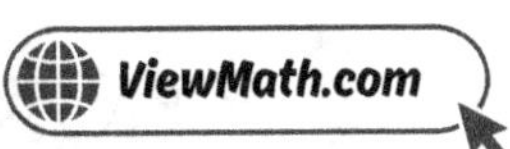

4. *A show starts at 5:30 and lasts 2 hours and 15 minutes. When does it end?* ________________

5. *How much time passes from 9:10 to 11:40?* ________________

6. *Lily starts reading at 6:15 P.M. and stops at 7:45 P.M. How long did she read?*

Answer: ________________ *time*

★ 5.3 Measuring Length ★

◎ In this lesson you will learn:

- Measure objects to the nearest $\frac{1}{2}$ inch and $\frac{1}{4}$ inch
- Use a ruler correctly starting at 0

🎓 Using a Ruler

A **ruler** measures how long something is. Always line up one end of the object with the **0** mark.

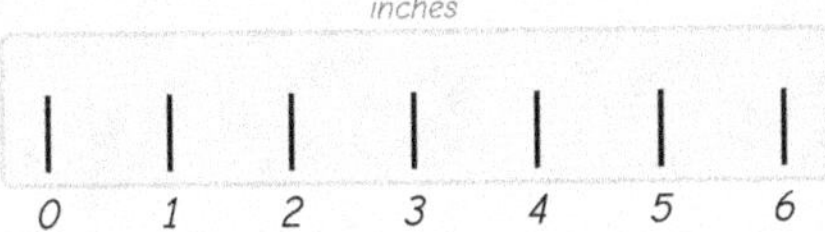

The lines between the whole numbers show **parts of an inch**:

- The **longest** middle line is the **half-inch** mark ($\frac{1}{2}$).
- The **shorter** lines split each half into **quarter inches** ($\frac{1}{4}$).

Between 0 and 1, the marks are:

$$0 \qquad \frac{1}{4} \qquad \frac{1}{2} \qquad \frac{3}{4} \qquad 1$$

“ Always start at **0**, not at 1! If you start at 1, your measurement will be 1 inch too big. ”

Measuring to the Nearest $\frac{1}{2}$ Inch

A crayon reaches past 3 inches and stops at the halfway mark before 4.

1. The crayon passed 3 inches.

2. It stopped at the half-inch mark between 3 and 4.

> ✔ **Answer:** $3\frac{1}{2}$ **inches**

Measuring to the Nearest $\frac{1}{4}$ Inch

A paper clip ends at the first small mark past 1 inch.

1. The paper clip passed 1 inch.

2. It lines up with the first quarter mark: $\frac{1}{4}$.

> ✔ **Answer:** $1\frac{1}{4}$ **inches**

✏ Measuring Length Practice ✏

1. A pencil reaches exactly to the 5-inch mark. How long is it? ___________

2. A ribbon ends at the halfway mark between 2 and 3. How long is the ribbon? ___________

3. An eraser ends at the third quarter mark past 1 inch. How long is it? ___________

4. A leaf reaches to the first quarter mark past 4 inches. How long is the leaf? ___________

5. A straw ends at the halfway mark between 5 and 6. How long is the straw? ___________

Find more at
ViewMath.com/Grade3

6. Emma measures two stickers. One is $2\frac{1}{4}$ inches long and the other is $1\frac{3}{4}$ inches long. How long are the two stickers placed end to end?

Answer: _________________ inches

⭐ 5.4 Mass and Weight ⭐

◎ In this lesson you will learn:

- *Measure mass in grams and kilograms*
- *Know that 1 kg = 1,000 g*
- *Solve word problems about mass*

🎓 Grams and Kilograms

Mass tells us how heavy an object is. We use two units:

- **Grams (g)** — for **light** things.
- **Kilograms (kg)** — for **heavy** things.

Key fact: 1 kilogram = 1,000 grams

About 1 **Gram**	About 1 **Kilogram**
A paper clip	A textbook
A dollar bill	A pineapple
A gummy bear	A bag of flour

Rule of thumb: If you can barely feel it in your hand, use **grams**. If it has real weight, use **kilograms**.

❝ *A paper clip is about 1 gram. A textbook is about 1 kilogram. Use these to help you guess the right unit!* ❞

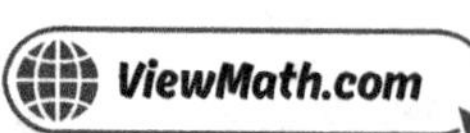

Adding Mass

A bag of apples has a mass of 3 kg. A bag of oranges has a mass of 2 kg. What is the total mass?

1. Apples: 3 kg. Oranges: 2 kg.

2. Total: $3 + 2 = 5$.

Answer: 5 **kg**

Comparing Mass

A melon has a mass of 2 kg. A bunch of grapes has a mass of 500 g. How much heavier is the melon in grams?

1. Convert the melon: $2 \text{ kg} = 2{,}000 \text{ g}$.

2. Subtract: $2{,}000 - 500 = 1{,}500$.

Answer: 1,500 **g**

✏ Mass and Weight Practice ✏

1. Would you measure a feather in grams or kilograms? __________

2. Would you measure a dog in grams or kilograms? __________

3. $3 \text{ kg} =$ __________ g

4. A box has a mass of 7 kg. You remove items with a mass of 3 kg. What mass is left? __________

5. $5{,}000 \text{ g} =$ __________ kg

6. Jake buys 4 bags of rice. Each bag has a mass of 2 kg. What is the total mass of all the bags?

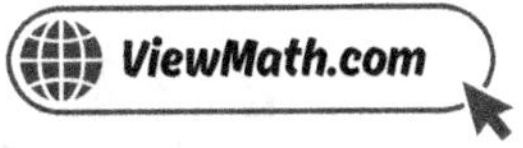

Answer: _________________ kg

★ 5.5 Liquid Volume & Capacity ★

◎ In this lesson you will learn:

- Measure liquid volume in liters (L) and milliliters (mL)
- Know that 1 L = 1,000 mL
- Solve word problems about liquid volume

🎓 Liters and Milliliters

Liquid volume tells us how much liquid a container can hold. We use two units:

- **Liters (L)** — for **large** amounts of liquid.
- **Milliliters (mL)** — for **small** amounts of liquid.

Key fact: 1 liter = 1,000 milliliters

About 1 **Milliliter**	About 1 **Liter**
A few drops of water	A water bottle
$\frac{1}{5}$ of a teaspoon	A large milk carton
One eyedropper squeeze	A big sports drink

Helpful benchmarks:

- A teaspoon holds about 5 mL.
- A drinking glass holds about 250 mL.
- A water bottle holds about 1 L.
- A bathtub holds about 150 L.

> 66 *Think of a regular water bottle — that's about 1 liter, which is 1,000 milliliters. A teaspoon is only about 5 mL — that's tiny!* 99

Adding Liquid Volumes

A fish tank has 12 L of water. You pour in 5 more liters. How much water is in the tank now?

1. Start with: 12 L.

2. Add: $12 + 5 = 17$.

> ✅ **Answer:** 17 **L**

Converting Liters to Milliliters

Sam has 3 L of juice. How many milliliters is that?

1. $1 L = 1,000$ mL.

2. $3 L = 3 \times 1,000 = 3,000$ mL.

> ✅ **Answer:** 3,000 **mL**

✏️ Liquid Volume Practice ✏️

1. Would you measure a spoonful of medicine in liters or milliliters? ___________

2. Would you measure water in a swimming pool in liters or milliliters? ___________

3. $2 L =$ ___________ mL

Find more at
ViewMath.com/Grade3

4. 5,000 mL = ___________ L

5. A pot holds 4 L of soup. You eat 1 L. How much soup is left? ___________

6. Maria fills 6 bottles with 1 L of water each. How many liters of water does she use in all?

Answer: ___________________ L

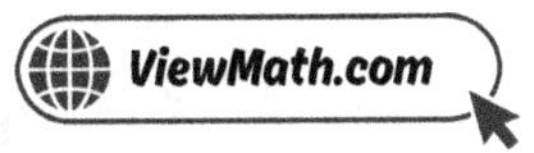

⭐ 5.6 Counting Money ⭐

◎ In this lesson you will learn:

- Identify coins and their values
- Count groups of coins and bills
- Make change from a purchase

🎓 Know Your Coins and Bills

Here are the coins you need to know:

Coin	Value	Remember
Penny	1c	1 cent
Nickel	5c	5 cents
Dime	10c	10 cents
Quarter	25c	25 cents

Key facts:

- 100 *cents* = $1.00.
- Always write **two digits** after the decimal point: $0.05, *not* $0.5.

Counting strategy: Start with the coin worth the **most** (quarters first), then dimes, nickels, and pennies last. **Skip count** by each coin's value.

❝ *A dime is worth MORE than a nickel even though it's smaller! Don't let the size fool you. Always count from the biggest value to the smallest.* **❞**

Count Mixed Coins

You have: 2 quarters, 1 dime, 1 nickel, and 3 pennies. How much money is that?

1. Quarters: $25, 50.$

2. Dime: $50 + 10 = 60.$

3. Nickel: $60 + 5 = 65.$

4. Pennies: $66, 67, 68.$

> ✓ **Answer:** 68 **cents, or** $0.68

Making Change

A sticker costs $0.65. You pay with $1.00. How much change do you get?

1. Amount paid: $1.00.

2. Price: $0.65.

3. Change: $1.00 − $0.65 = $0.35.

> ✓ **Answer:** $0.35

Counting Money Practice

1. 3 dimes and 2 pennies = ____________

2. 1 quarter and 3 nickels = ____________

3. 2 quarters, 1 dime, 4 pennies = ____________

4. 45 cents = $ __________

5. Price: $0.75. You pay $1.00. Change: ____________

6. Emma has 3 quarters and 2 dimes. How much money does she have?

Answer: ________________

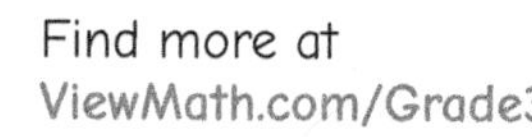
Find more at
ViewMath.com/Grade3

★ 5.7 Adding & Subtracting Money ★

◎ In this lesson you will learn:

- Add and subtract money amounts using decimal notation
- Line up decimal points before computing
- Make change by subtracting

🎓 Adding and Subtracting Money

Adding and subtracting money works just like regular addition and subtraction. Follow these steps:

1. Write the amounts with the **decimal points lined up**.
2. Add or subtract the **cents** first (digits to the right of the dot).
3. Then add or subtract the **dollars** (digits to the left of the dot).
4. Regroup (carry or borrow) if needed.
5. Put the **$ sign** and **decimal point** in your answer!

Tip: Write $5 as $5.00 so the decimal points line up evenly.

Making change means finding out how much money you get back:

$$\textbf{Change} = \text{Amount Paid} - \text{Price}$$

> 66 Always line up the decimal points before you add or subtract! If the dots aren't stacked, your answer will be wrong. 99

Adding Money

Add $3.45 + $2.30.

$$\begin{array}{r} \$3.45 \\ +\ \$2.30 \\ \hline \$5.75 \end{array}$$

1. Cents: $5 + 0 = 5$.

2. Tens of cents: $4 + 3 = 7$.

3. Dollars: $3 + 2 = 5$.

✓ **Answer:** $5.75

Making Change

You buy a book for $6.75. You pay with $10.00. How much change do you get?

$$\begin{array}{r} \$10.00 \\ -\ \$\ 6.75 \\ \hline \$\ 3.25 \end{array}$$

1. Can't take 5 from 0 — borrow!

2. Cents: $10 - 5 = 5$.

3. Tens of cents: $9 - 7 = 2$ (after borrowing).

4. Dollars: $9 - 6 = 3$ (after borrowing).

✓ **Answer:** $3.25

Find more at
ViewMath.com/Grade3

Adding & Subtracting Money Practice

1. $2.50 + $1.25 = _______________

2. $4.30 + $3.60 = _______________

3. $8.50 − $3.25 = _______________

4. $10.00 − $4.60 = _______________

5. A pen costs $1.50 and a folder costs $2.25. What is the total cost? _______________

6. Luke has $12.00. He buys a hat for $7.50. How much money does Luke have left?

Answer: _______________

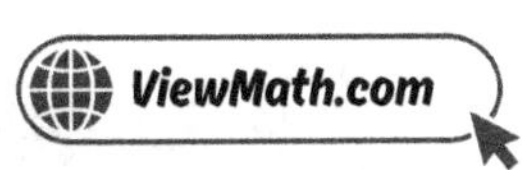

★ 5.8 Personal Financial Literacy ★

◎ In this lesson you will learn:

- Tell the difference between **needs** and **wants**
- Understand what a **budget** is and use the three-jar method

🎓 Needs, Wants, and Budgets

Needs are things you must have to live: food, water, shelter, clothing.

Wants are extras that are nice to have: toys, candy, video games.

Needs always come first! After needs are met, you can think about wants.

A **budget** is a plan for your money. It tells every dollar where to go.

The Three-Jar Method:

SAVE	SPEND	SHARE
Keep for later	Use for things	Give to help
(bigger goals!)	you want now	others

If you get $12, you might put: $6 in Save, $4 in Spend, $2 in Share.

66 Before you buy something, ask: "Do I **need** it or do I **want** it?" This helps you make smart money choices! 99

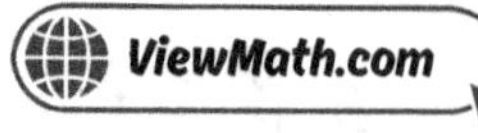

Making a Budget

Jake has $10. He puts half in Save and splits the rest between Spend and Share.

1. *Save:* $\$10 \div 2 = \$5$

2. *Remaining:* $\$10 - \$5 = \$5$

3. *Spend:* $\$5 \div 2 = \$2.50$ *Share:* $\$2.50$

> ✔ **Answer: Save:** $\$5$**, Spend:** $\$2.50$**, Share:** $\$2.50$

Need or Want?

Mia has $15. She needs school supplies for $9 and wants a toy for $8. Can she buy both?

1. *Total cost:* $\$9 + \$8 = \$17$

2. *She only has $15, so $17 > $15 — not enough!*

3. *Needs come first: buy the school supplies ($9)*

4. *Money left:* $\$15 - \$9 = \$6$ *(not enough for the $8 toy)*

> ✔ **Answer: No — she should buy the school supplies (need) first. She has** $\$6$ **left, which is not enough for the toy.**

✏ Financial Literacy Practice ✏

1. Is food a **need** or a **want?** ____________

2. Is a video game a **need** or a **want?** ____________

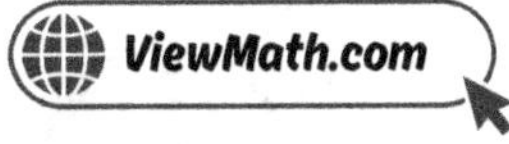

3. Sam has $20. He saves half. How much does he save? _____________

4. Lily has $15. She spends $7 on lunch and $3 on a book. How much money does she have left? _____________

5. Marco gets $9. He puts $3 in Save, $4 in Spend, and the rest in Share. How much goes in Share? _____________

6. Emma has $18. She needs new shoes for $12. She wants a toy for $10. Can she buy both? If not, what should she buy first?

Answer: _____________

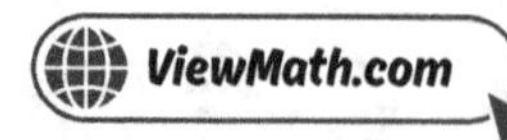

6

Data & Graphs

What's Inside

★ 6.1 Picture Graphs ★

◉ In this lesson you will learn:

- Read picture graphs where each symbol stands for more than 1 item
- Answer "how many more" and "how many fewer" questions

🎓 Reading Picture Graphs

A **picture graph** (or **pictograph**) uses symbols to show data. The **key** at the bottom tells you what each symbol stands for.

📊 Shells Collected

Mia ★ ★ ★ ★ ★

Jake ★ ★ ★

Lily ★ ★ ★ ★ ★ ★ ★

Sam ★ ★ ★ ★

Key: Each ★ = 2 shells

Each ★ stands for 2 shells. To find the total, **multiply**:

- Mia: $5 \times 2 = 10$ shells
- Jake: $3 \times 2 = 6$ shells
- Lily: $7 \times 2 = 14$ shells
- Sam: $4 \times 2 = 8$ shells

Find more at
ViewMath.com/Grade3

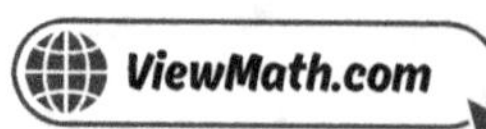

How many more shells did Lily collect than Jake?

$14 - 6 = 8$ more shells.

How many shells in all?

$10 + 6 + 14 + 8 = 38$ shells.

 66 *Always read the* **key** *first! If each symbol* $= 5$, *then 4 symbols means* $4 \times 5 = 20$, *not just 4!* 99

Reading a Picture Graph

📊 Books Read This Month

Emma 💜💜💜💜

Noah 💜💜💜💜💜💜

Ava 💜💜💜

Liam 💜💜💜💜💜

Key: Each 💜 = 3 books

Each ♥ stands for 3 books.

1. Emma: $4 \times 3 = 12$ books

2. Noah: $6 \times 3 = 18$ books

3. Ava: $3 \times 3 = 9$ books

4. Liam: $5 \times 3 = 15$ books

5. How many more books did Noah read than Ava? $18 - 9 = 9$ more books

> ✓ **Answer:** Noah read 9 more books than Ava.

✏ Picture Graphs Practice ✏

Use this picture graph to answer Problems 1–6.

Favorite Ice Cream Flavors

Chocolate ★ ★ ★ ★ ★ ★

Vanilla ★ ★ ★ ★

Strawberry ★ ★ ★

Mint ★ ★ ★ ★ ★

Key: Each ★ = 2 votes

1. How many votes did Chocolate get? ___________

2. How many votes did Strawberry get? ___________

3. Which flavor got the most votes? ___________

4. How many more votes did Mint get than Strawberry? ___________

5. How many votes were there in all? ___________

6. In a picture graph, each ★ stands for 5 stickers. The row for Monday has 4 stars and the row for Tuesday has 7 stars. How many more stickers were given out on Tuesday than Monday?

Answer: ___________ stickers

Find more at
ViewMath.com/Grade3

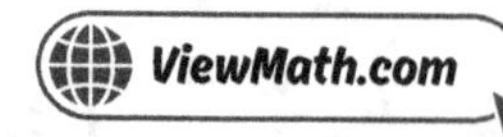
ViewMath.com

★ 6.2 Bar Graphs ★

◎ In this lesson you will learn:

- Read bar graphs using the scale
- Compare data and solve "how many more/fewer" problems

🎓 Reading Bar Graphs

A **bar graph** uses bars to show data. The **scale** on the side tells you the numbers. The taller the bar, the bigger the value!

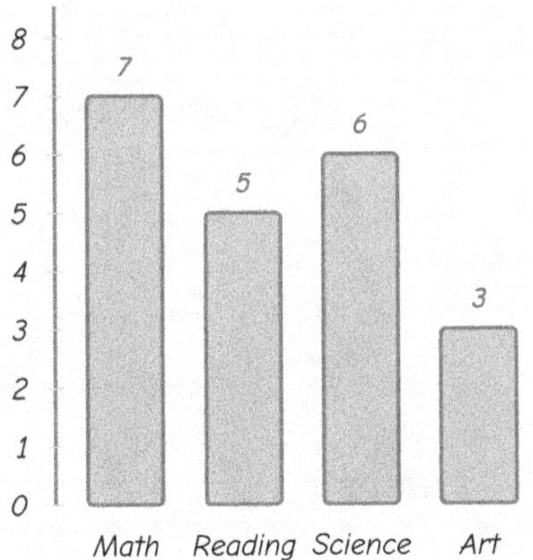

To read a bar graph:

- **Read the title** — it tells you what the graph is about.
- **Check the scale** — the numbers on the side.
- **Read each bar** — see where the top lines up with the scale.

Math has the tallest bar (7), so it is the most popular.

Art has the shortest bar (3), so it is the least popular.

How many more students chose Math than Art?

$7 - 3 = 4$ more students.

How many students in all?

$7 + 5 + 6 + 3 = 21$ students.

❝ To find "how many more" or "how many fewer," just **subtract**! Find the two bars, read their values, and take the difference. ❞

Reading a Bar Graph

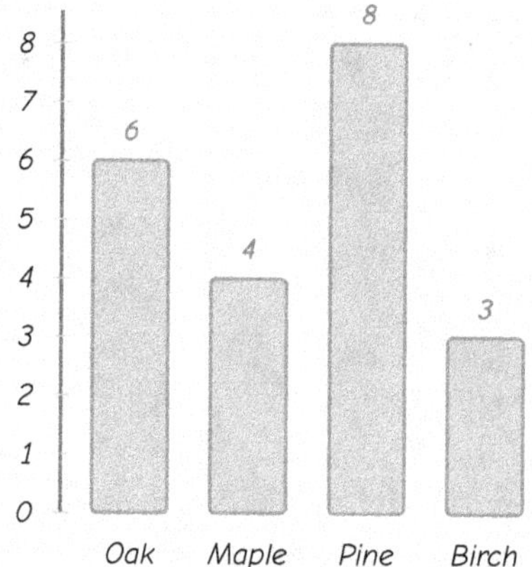

1. Which tree is there the most of? Pine — tallest bar (8).

2. Which tree is there the fewest of? Birch — shortest bar (3).

3. How many more Pine than Maple? $8 - 4 = 4$ more.

4. How many trees in all? $6 + 4 + 8 + 3 = 21$ trees.

> ✅ **Answer:** 21 **trees in all**

Find more at
ViewMath.com/Grade3

✎ Bar Graphs Practice ✎

Use this bar graph to answer Problems 1–5.

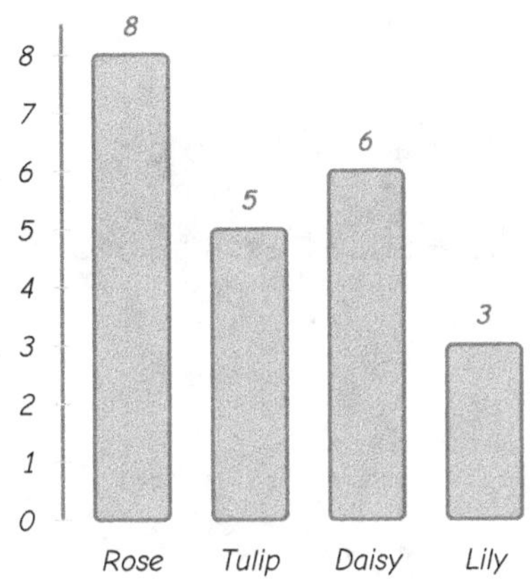

1. How many roses are in the garden? ___________

2. Which flower has the fewest? ___________

3. How many more roses than lilies? ___________

4. How many tulips and daisies combined? ___________

5. How many flowers are there in all? ___________

6. A bar graph shows that Room A has 12 students, Room B has 8 students, and Room C has 15 students. How many more students are in Room C than Room B?

Answer: ___________________ students

⭐ 6.3 Line Plots ⭐

◎ In this lesson you will learn:

- Read and make line plots with measurement data
- Use line plots to answer questions about data

🎓 What Is a Line Plot?

A **line plot** is a number line with **X marks** above it. Each X stands for one piece of data. X marks are stacked when values repeat.

A class measured the lengths of 12 crayons to the nearest inch. Here are the results shown on a line plot:

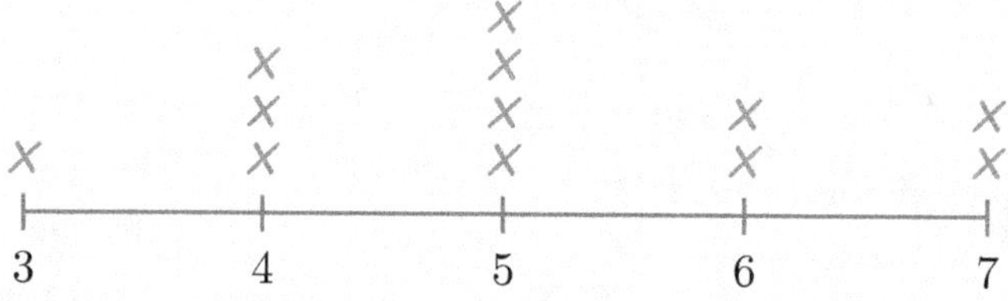

Count the X marks above each number to read the data:

- 3 inches: 1 crayon   •   4 inches: 3 crayons   •   5 inches: 4 crayons
- 6 inches: 2 crayons   •   7 inches: 2 crayons

Most common length: 5 inches (the most X marks — 4!).

Total crayons: $1 + 3 + 4 + 2 + 2 = 12$.

> **"** Line plots are great for **measurement data**! Measure lengths, heights, or widths, then stack X marks to see which measurement shows up the most. **"**

Reading a Line Plot

Students measured leaves and made this line plot:

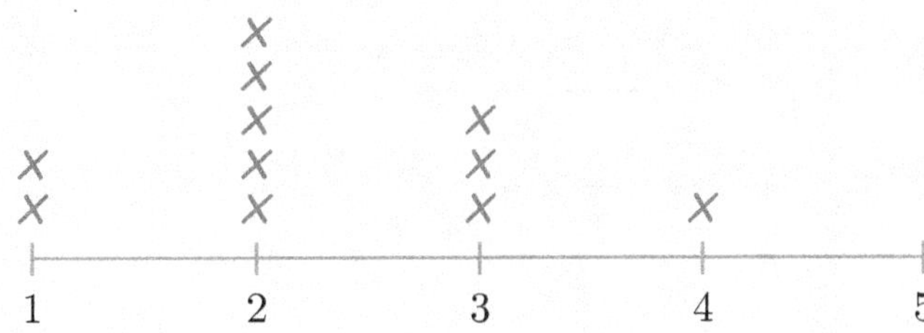

1. How many leaves are 2 inches long? Count the X marks above 2: there are 5.

2. What is the most common length? 2 inches — it has the most X marks.

3. How many leaves were measured in all? $2 + 5 + 3 + 1 + 0 = 11$ leaves.

4. How many more 2-inch leaves than 4-inch leaves? $5 - 1 = 4$ more.

> **Answer:** **11 leaves were measured in all.**

Line Plots Practice

Use this line plot to answer Problems 1–6. It shows the lengths of ribbons (in inches) that students cut in art class.

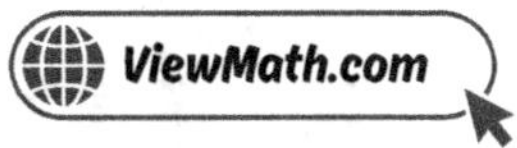

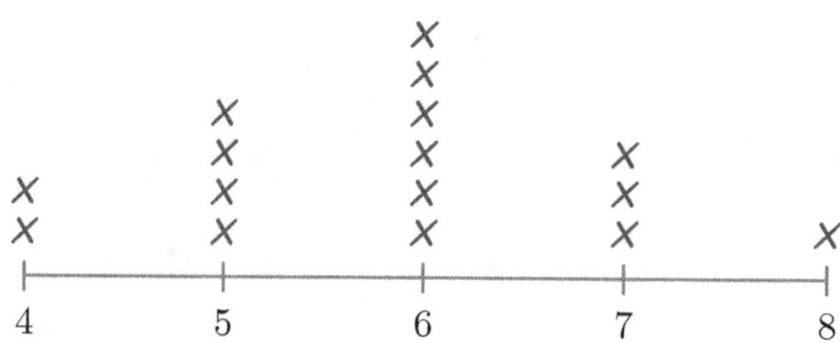

1. How many ribbons are 6 inches long? __________

2. What is the most common ribbon length? __________

3. How many ribbons are 4 inches long? __________

4. How many more 6-inch ribbons than 8-inch ribbons? __________

5. How many ribbons were measured in all? __________

6. Look at the line plot. How many ribbons are 5 inches or shorter?

Answer: __________________ *ribbons*

Find more at
ViewMath.com/Grade3

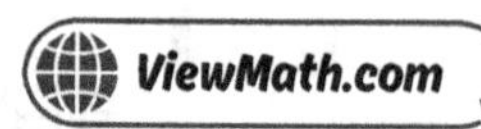

CHAPTER

7

Geometry

What's Inside

★ 7.1 2D Shapes & Attributes ★

◎ In this lesson you will learn:

- *Identify quadrilaterals by their sides, angles, and parallel sides*
- *Understand that a square is a special rectangle AND a special rhombus*

🎓 The Quadrilateral Family

A **quadrilateral** is any shape with exactly 4 sides. Here are the most important types:

| Square | Rectangle | Rhombus | Trapezoid |

Shape	Sides	Angles	Parallel Sides
Square	4 equal	4 right	2 pairs
Rectangle	2 pairs equal	4 right	2 pairs
Rhombus	4 equal	not always right	2 pairs
Trapezoid	varies	varies	1 pair

A **right angle** looks like the corner of a piece of paper.

Parallel sides go in the same direction and never cross, like railroad tracks.

Key idea: Every **square** is a rectangle (it has 4 right angles). Every square is also a rhombus (it has 4 equal sides). But not every rectangle is a square, and not every rhombus is a square!

> *Think of a square as the "perfect" quadrilateral — it has equal sides AND right angles. It belongs to many shape families!*

Classify a Shape

A shape has 4 sides. Two sides are 8 cm and two sides are 5 cm. All 4 angles are right angles. What shape is it?

1. It has 4 sides, so it is a **quadrilateral**.

2. It has 4 right angles — that means it could be a rectangle or a square.

3. Are all 4 sides equal? No (8 cm and 5 cm) — so it is NOT a square.

> ✓ **Answer: It is a rectangle.**

✏ 2D Shapes Practice ✏

1. I have 4 equal sides and 4 right angles. What shape am I? ________________

2. I have 4 sides and only 1 pair of parallel sides. What shape am I? ________________

3. I have 4 equal sides but my angles are NOT all right angles. What shape am I? ________________

4. A square is a type of rectangle.   True ○   False ○

5. All rectangles are squares.   True ○   False ○

6. A garden has 4 right angles and 2 pairs of equal sides, but not all sides are the same length. What shape is the garden?

Answer: ________________

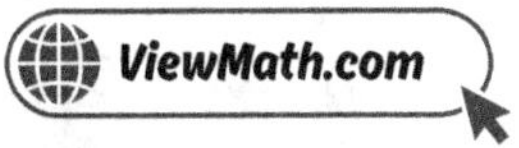

★ 7.2 3D Shapes: Prisms & Pyramids ★

◎ In this lesson you will learn:

- Identify 3D shapes including **prisms** and **pyramids**
- Classify 3D shapes by their faces, edges, and vertices

🎓 3D Shapes: Prisms, Pyramids, and More

3D means **three-dimensional**. A 3D shape is solid — you can hold it in your hand! Every 3D shape can have:

- **Faces** — the flat surfaces on the outside
- **Edges** — the lines where two faces meet
- **Vertices** — the corners where edges meet

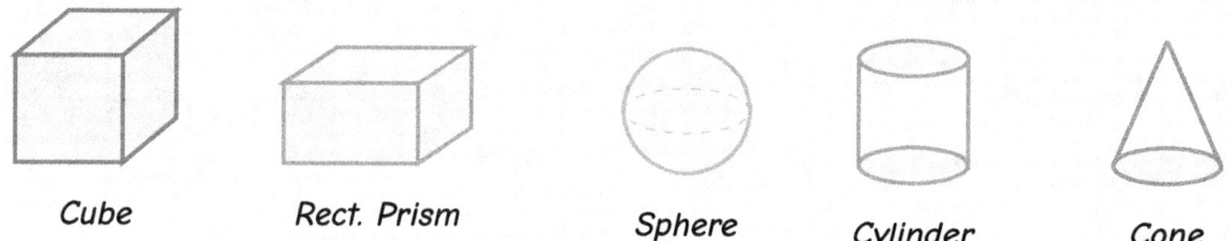

A **prism** has two identical flat faces called **bases**, connected by rectangles. You name a prism by the shape of its base. A **cube** is a special prism where every face is a square. A **rectangular prism** has rectangle faces.

A **pyramid** has one flat **base** and triangular faces that meet at a point called the **apex**. You name a pyramid by the shape of its base.

A **sphere** has NO faces, NO edges, NO vertices — it is perfectly round!

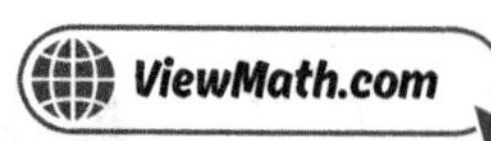

Shape	Faces	Edges	Vertices
Cube	6	12	8
Rectangular Prism	6	12	8
Triangular Prism	5	9	6
Square Pyramid	5	8	5
Triangular Pyramid	4	6	4
Sphere	0	0	0
Cylinder	2 flat	2	0
Cone	1 flat	1	1

66 *A cube is like a square's 3D cousin! The **faces** of 3D shapes ARE 2D shapes — a cube's faces are squares, and a cylinder's flat faces are circles.* **99**

Describe a Rectangular Prism

How many faces, edges, and vertices does a rectangular prism have?

1. Count the flat surfaces: 6 faces (top, bottom, front, back, left, right).

2. Count the lines where faces meet: 12 edges.

3. Count the corners: 8 vertices.

✓ **Answer: A rectangular prism has** 6 **faces,** 12 **edges, and** 8 **vertices.**

Describe a Square Pyramid

How many faces, edges, and vertices does a square pyramid have?

1. The base is a square — that is 1 face.

2. It has 4 triangular faces that rise up to the apex — that gives 4 more faces. Total: 5 faces.

3. Count the edges: the base has 4 edges, and 4 more edges go from each base corner up to the apex. Total: 8 edges.

4. Count the vertices: the base has 4 corners, plus 1 apex at the top. Total: 5 vertices.

> ✓ **Answer: A square pyramid has 5 faces, 8 edges, and 5 vertices.**

3D Shapes Practice

1. I have 6 square faces, 12 edges, and 8 vertices. What am I? _______________

2. I have no faces, no edges, and no vertices. I am perfectly round. What am I? _______________

3. I have 2 triangular bases and 3 rectangular faces. What am I? _______________

4. A soup can is shaped like a _______________.

5. How many faces does a triangular pyramid have? _______________

6. The Great Pyramid of Egypt has a square base and 4 triangular faces. How many edges and vertices does it have?

Answer: _______________

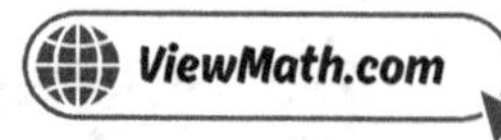

⭐ 7.3 What Is Area? ⭐

◎ In this lesson you will learn:

- Understand that area is the space inside a flat shape
- Measure area by counting unit squares

🎓 Area = Counting Unit Squares

Area is the amount of space a flat shape covers. We measure area by counting how many **unit squares** fit inside the shape.

Look at this rectangle covered with unit squares:

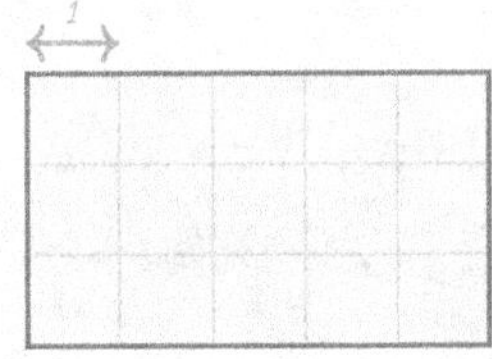

Count the squares: 3 rows × 5 columns = 15 unit squares.
The area is **15 square units**!

A **unit square** is a square with sides that are 1 unit long. Area is always measured in **square units** (sq units, sq cm, sq in).

Two ways to find area:

- **Count** every unit square inside the shape.
- **Multiply:** rows × columns (faster for rectangles!).

> *Think of area like tiling a floor! How many square tiles fit inside the shape? Count them all up and you have the area!*

Find the Area by Counting

Count the unit squares to find the area.

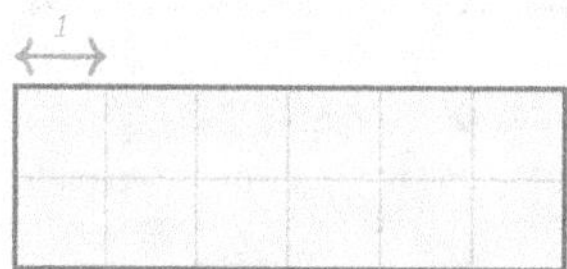

1. Row 1 has 6 squares. Row 2 has 6 squares.

2. Total: $6 + 6 = 12$ squares.

3. Or multiply: 2 rows $\times$ 6 columns $= 12$.

✓ **Answer: Area =** 12 **square units**

✏ Area Practice ✏

1. Find the area.

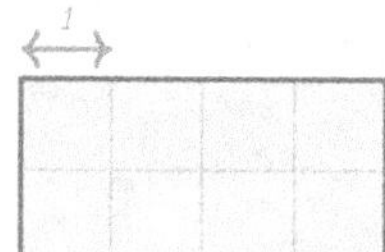

Area = ______________

2. Find the area.

Find more at
ViewMath.com/Grade3

Area = ___________________

3. *A rectangle has 4 rows and 7 columns of unit squares. What is the area?* ___________________

4. *A rectangle has 5 rows and 5 columns of unit squares. What is the area?* ___________________

5. *A 3 × 5 rectangle has an area of 15 square units.*   **True** ○   **False** ○

6. *Sam's garden is 6 feet long and 3 feet wide. What is the area of the garden?*

Answer: ___________________ *square feet*

⭐ 7.4 Finding Area ⭐

◉ In this lesson you will learn:

- Use the formula: Area = length × width
- Split complex shapes into rectangles and add the areas

🎓 Area = Length × Width

Instead of counting every square, use the **area formula**:

$$Area = length \times width$$

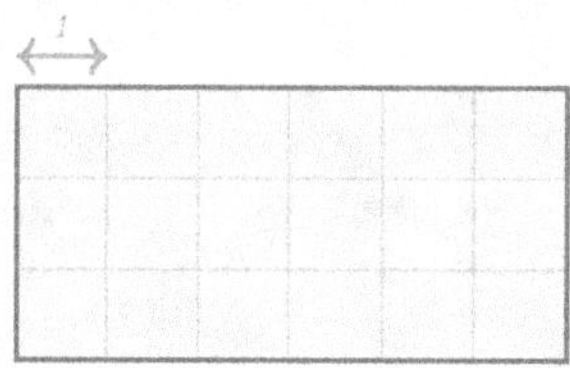

3 rows × 6 columns = 18 square units. That's the same as length × width!

For complex shapes: Split the shape into rectangles, find each area, then **add** them together.

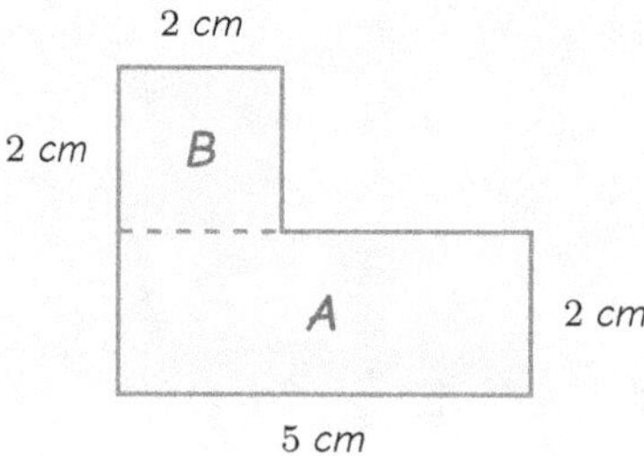

Area of A = 5 × 2 = 10 sq cm.   Area of B = 2 × 2 = 4 sq cm.

Total area = 10 + 4 = **14 sq cm.**

Find more at
ViewMath.com/Grade3

> *Remember: area is always in **square** units — sq cm, sq in, sq ft. Multiply length times width and you're done!*

Find the Area of a Rectangle

A rectangle is 7 inches long and 3 inches wide. What is its area?

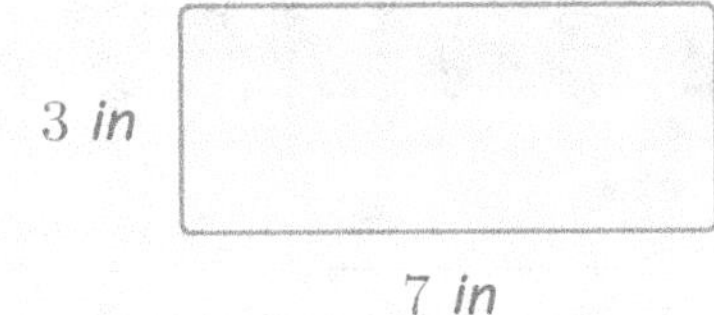

1. Write the formula: Area = length × width

2. Plug in the numbers: Area = 7×3

3. Multiply: $7 \times 3 = 21$

✓ **Answer: Area =** 21 **square inches**

Split and Add

Find the total area of this T-shape.

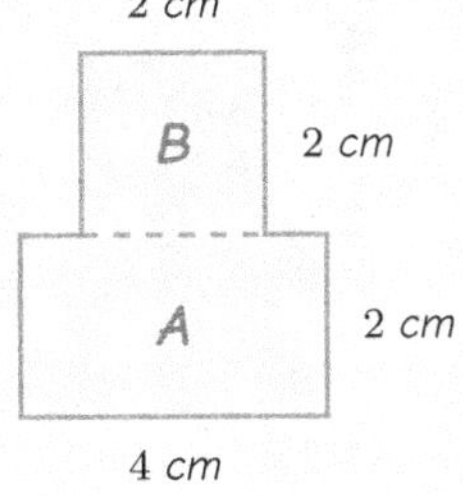

1. Area of A (bottom): $4 \times 2 = 8$ sq cm

2. Area of B (top): $2 \times 2 = 4$ sq cm

Find more at
ViewMath.com/Grade3

ViewMath.com

3. *Total area:* $8 + 4 = 12$ *sq cm*

> ✔ **Answer:** **Total area =**   **12** *square cm*

✏ Finding Area Practice ✏

1. *Length = 8 cm, Width = 3 cm.*   *Area =* ___________________

2. *Length = 5 in, Width = 5 in.*   *Area =* ___________________

3. *Length = 9 ft, Width = 4 ft.*   *Area =* ___________________

4. *Find the area.*

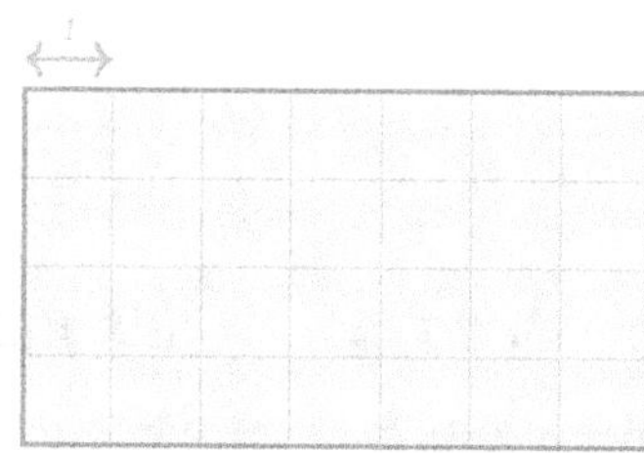

Area = ___________________

5. *Find the total area of this L-shape.*

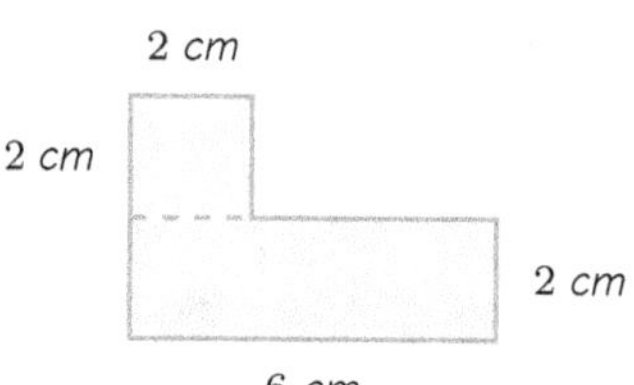

Total area = ___________________

6. *A classroom rug is 6 feet long and 4 feet wide. What is the area of the rug?*

Answer: ___________________ *square feet*

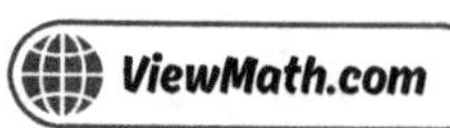

★ 7.5 Perimeter ★

◎ *In this lesson you will learn:*

- *Find the perimeter of shapes by adding all side lengths*
- *Find a missing side when the perimeter is known*
- *Understand that same perimeter does not mean same area*

🎓 Perimeter = Distance Around

Perimeter is the total distance around the outside of a shape. Add up **all the side lengths** to find it.

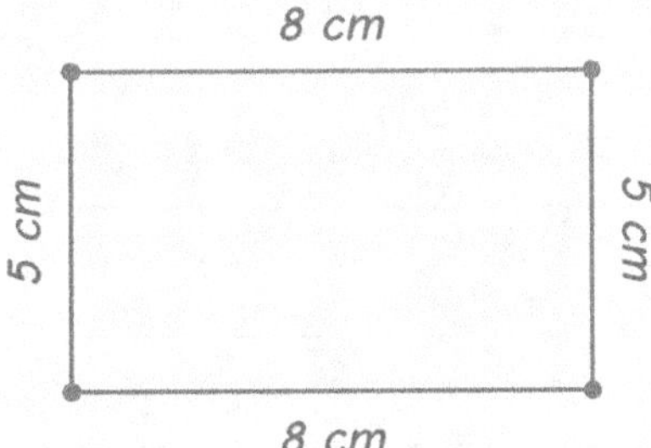

The perimeter is $8 + 5 + 8 + 5 = $ **26 cm**.

Rectangle shortcut:

$$Perimeter = (2 \times length) + (2 \times width)$$

$(2 \times 8) + (2 \times 5) = 16 + 10 = 26$ cm ✓

Finding a missing side: If you know the perimeter and all sides except one, subtract to find it!

Example: Perimeter = 20 cm, length = 7 cm. Find the width.

$20 = 2 \times 7 + 2 \times width$, so $2 \times width = 20 - 14 = 6$, width = 3 cm.

Same perimeter ≠ same area! A 7×1 rectangle and a 4×4 square both have perimeter = 16 cm, but their areas are 7 sq cm and 16 sq cm — very different!

Find more at
ViewMath.com/Grade3

> *Think of perimeter like building a fence around a garden. How much fence do you need? Add up all 4 sides!*

Find the Perimeter of a Rectangle

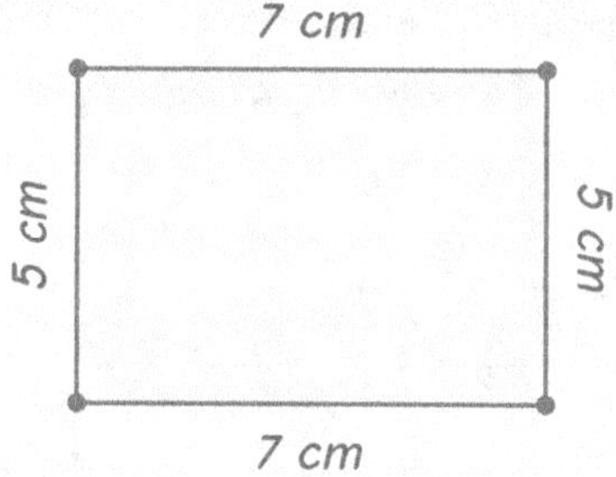

1. Write all four sides: $7, 5, 7, 5$

2. Add them up: $7 + 5 + 7 + 5 = 24$

✔ **Answer: Perimeter** $=$ **24 cm**

Find a Missing Side

A rectangle has a perimeter of 22 cm and a length of 7 cm. What is the width?

1. Use the formula: $22 = (2 \times 7) + (2 \times width)$

2. Simplify: $22 = 14 + 2 \times width$

3. Subtract: $2 \times width = 22 - 14 = 8$

4. Divide: $width = 8 \div 2 = 4$ cm

✔ **Answer: Width** $=$ **4 cm**

Find more at
ViewMath.com/Grade3

Perimeter Practice

1. *Find the perimeter.*

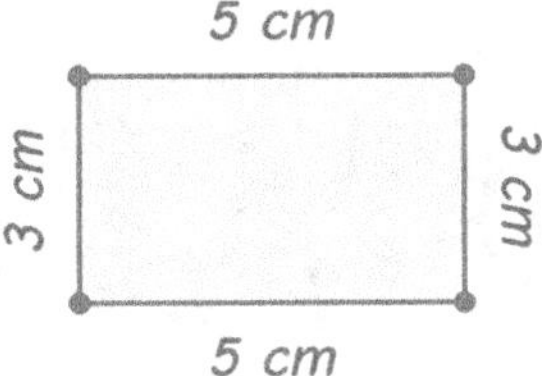

Perimeter = ______________

2. *Find the perimeter.*

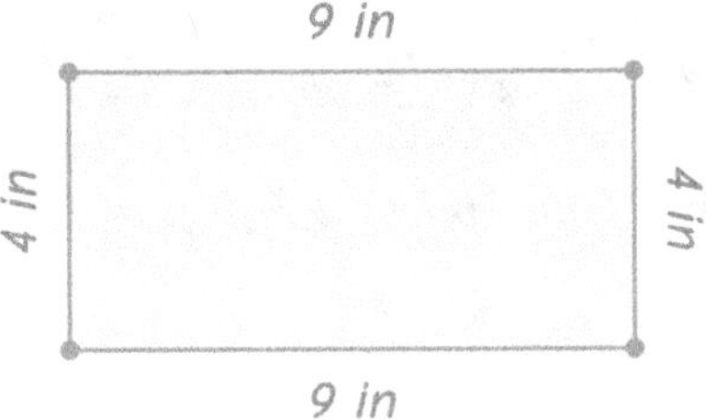

Perimeter = ______________

3. A square has sides of 6 cm.   Perimeter = ______________

4. A rectangle has a perimeter of 30 in and a width of 5 in.   Length = ______________

5. A triangle has sides of 5 cm, 4 cm, and 3 cm.   Perimeter = ______________

6. A garden is 8 meters long and 5 meters wide. How much fencing is needed to go all the way around it?

Answer. ______________ meters

★ 7.6 Partitioning Shapes ★

◎ In this lesson you will learn:

- Partition shapes into equal parts and name each part with a unit fraction
- Understand that equal parts must be the same size

🎓 Partitioning = Making Equal Parts

To **partition** a shape means to divide it into **equal parts**. Each part is a **unit fraction** of the whole.

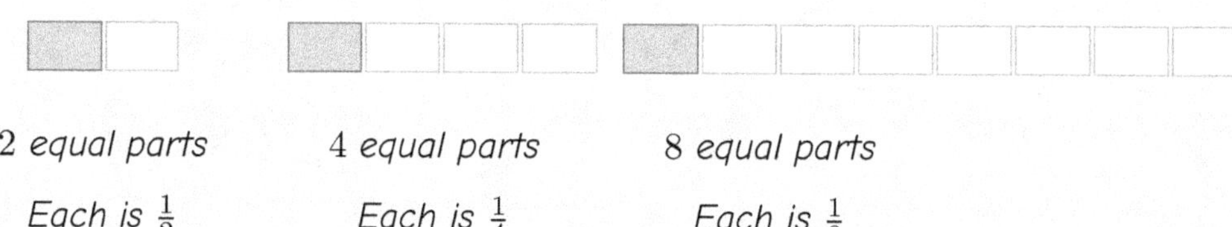

2 equal parts   4 equal parts   8 equal parts

Each is $\frac{1}{2}$ Each is $\frac{1}{4}$ Each is $\frac{1}{8}$

Circles work the same way:

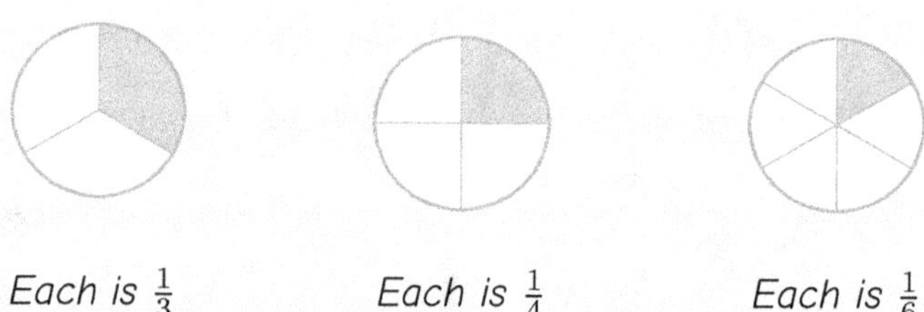

Each is $\frac{1}{3}$ Each is $\frac{1}{4}$ Each is $\frac{1}{6}$

Key rule: The parts must be **equal in size**. If the parts are different sizes, you cannot use fractions to name them!

More equal parts = **smaller** pieces. $\frac{1}{8}$ is smaller than $\frac{1}{4}$ because you are splitting the same shape into more pieces.

> *Partitioning is where fractions come from! Every fraction you learned started by cutting a shape into equal pieces.*

Partition and Label

A square is cut into 4 equal parts. What fraction is each part?

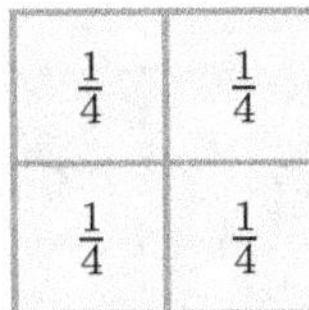

1. Count the equal parts: 4

2. Write the unit fraction: $\frac{1}{4}$

> ✅ **Answer:** **Each part** $= \quad \frac{1}{4}$

Equal or Unequal?

A rectangle is split into 3 parts. Are they equal? Can we use fractions?

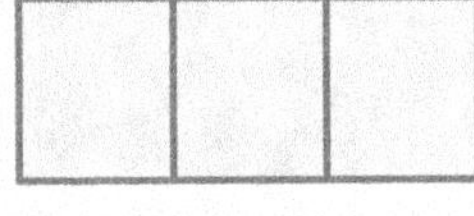
Equal ✓

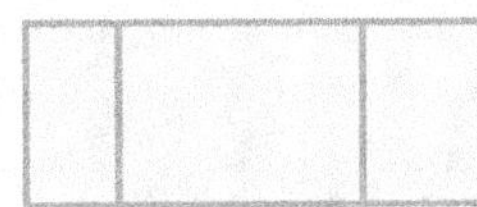
Unequal X

1. The blue rectangle has 3 parts that are the **same size** — each part is $\frac{1}{3}$ ✓

2. The red rectangle has 3 parts that are **different sizes** — we cannot name them with fractions!

> ✅ **Answer:** **Only equal parts can be named with fractions**

✏ Partitioning Practice ✏

1. Each part = __________

2. Each part = __________

3. Each part = __________

4. A shape is partitioned so each part is $\frac{1}{8}$. How many equal parts does it have? __________

5. Which is larger: $\frac{1}{4}$ of a rectangle or $\frac{1}{6}$ of the same rectangle? __________

6. A pizza is cut into 6 equal slices. You eat 2 slices. What fraction of the pizza did you eat?

Answer: __________________

Get Online

Find more at
ViewMath.com/Grade3

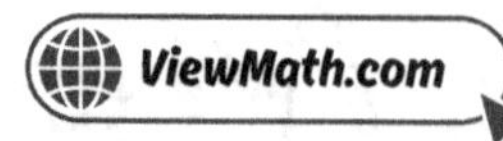

⭐ 7.7 Lines, Line Segments, Rays, and Angles ⭐

◎ In this lesson you will learn:

- Identify **lines**, **line segments**, and **rays**
- Understand what an **angle** is and classify angles as right, acute, or obtuse

🎓 Straight Paths and Angles

There are three types of straight paths in geometry:

Line   ←●————————●→   No endpoints — goes forever both ways

Segment   ●————————●   TWO endpoints — stops at both ends

Ray   ●————————→   ONE endpoint — goes forever one way

An **angle** is formed when two rays share the same endpoint, called the **vertex**. Angles are measured in **degrees** (°).

Angle Type	Size	Looks Like
Right angle	Exactly 90°	A perfect square corner
Acute angle	Less than 90°	Smaller than a square corner
Obtuse angle	Greater than 90°	Bigger than a square corner

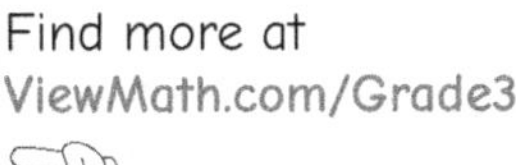

*The corner of a piece of paper is a **right angle** (90°). Hold it up to any angle to check — if the angle is smaller, it's acute; if it's bigger, it's obtuse!*

Identify the Straight Path

A flashlight beam starts at the bulb and shines forward. What type of straight path is it?

1. It has ONE starting point (the bulb) — that's an endpoint.

2. The light goes on and on in one direction.

3. One endpoint + goes forever one way = a **ray**.

> ✅ **Answer:** *A flashlight beam is a ray.*

Classify the Angle

The hands of a clock at 3:00 form what type of angle?

1. At 3:00, the minute hand points to 12 and the hour hand points to 3.

2. The opening between them makes a perfect square corner.

3. A square corner is exactly 90°.

> ✅ **Answer:** *The angle at* 3:00 *is a right angle.*

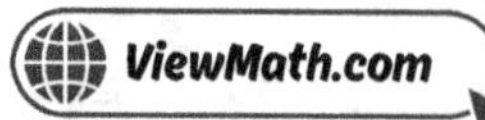

Lines, Segments, Rays, and Angles Practice

1. A ruler has a mark at each end. The edge between the two marks is what type of straight path?
____

2. Railroad tracks go on and on in both directions with no endpoints. What type of straight path?
____

3. A sunbeam starts at the sun and shoots outward. What type of straight path? _______________

4. An angle that measures 45° is what type of angle? _______________

5. An angle that measures 120° is what type of angle? _______________

6. How many endpoints does a line segment have? __________

Answer Key

 **Check Your Answers!** 

Try each problem first, then look here to check your work.
It's OK to make mistakes — that's how we learn ★

📖 Chapter 1

> Section 1.1

1 300 **2** 60 **3** $200 + 50 + 9$ **4** 639 **5** $400 + 0 + 6$ **6** \$352

> Section 1.2

1 $5,206 = 5,000 + 200 + 0 + 6$ **2** 9,000 **3** 7,461 **4** $2,856 < 2,912$ **5** $6,450 > 6,405$

6 $3,120 < 3,201 < 3,210$

> Section 1.3

1 $<$ **2** $>$ **3** $=$ **4** $<$ **5** $197, 284, 342$ **6** $5,021, \ 5,120, \ 5,210$

> Section 1.4

1 30 **2** 70 **3** 90 **4** 300 **5** 600 **6** 800

Find more at
ViewMath.com/Grade3

> Section 1.5

 E    O    E   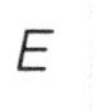 Odd    14, Even    11, Odd

> Section 1.6

 10    1    10,000    10,000    8,000, 10,000    10

> Section 1.7

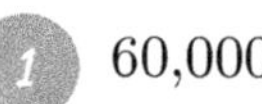 60,000    8,000    10    100    $40,000 + 5,000 + 900 + 0 + 8$

 6; *its value is* 60,000

> Section 1.8

 500,000    80,000    $200,000 + 50,000 + 6,000 + 300 + 70 + 4$ 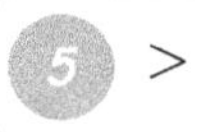 743,218   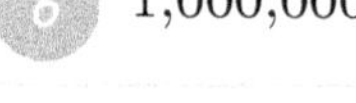 >

 1,000,000

Chapter 2

> Section 2.1

 557    557    834    436    852    623

> *Section 2.2*

 444  441  157  265  343  368

> *Section 2.3*

 4,686  6,664  6,245  6,243  8,542  8,145

> *Section 2.4*

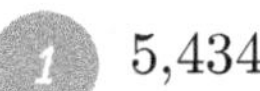 5,434  2,556  3,545 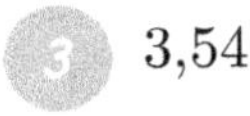 3,267  2,654  2,763

> *Section 2.5*

 800  1,000  500  400  *No*  *About* 800

📖 Chapter 3

> *Section 3.1*

 $3 \times 5 = 15$  $2 \times 7 = 14$  $24 \longrightarrow 4 \times 6 = 24$  15  16  20 *apples*

> *Section 3.2*

 0  9  16  30  28  27  20 *desks*

Find more at
ViewMath.com/Grade3

  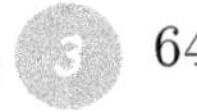    

Section 3.3

1 42　**2** 54　**3** 64　**4** 63　**5** 70　**6** 48　**7** 56 *books*

Section 3.4

1 70　**2** 400　**3** 400　**4** 270　**5** 1,000　**6** 160 *seats*

Section 3.5

  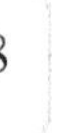 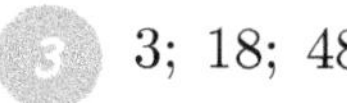  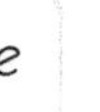  

1 5　**2** 28　**3** 3; 18; 48　**4** True　**5** False　**6** $35 + 7 = 42$

Section 3.6

1 5　**2** 6　**3** 7　**4** 4　**5** True　**6** 8 *students*

Section 3.7

   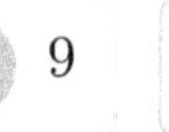   

1 9　**2** 8　**3** 9　**4** 7　**5** 9　**6** 6 *boxes*

Section 3.8

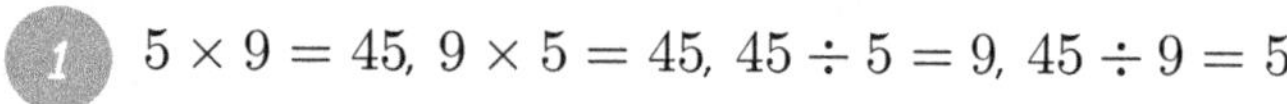  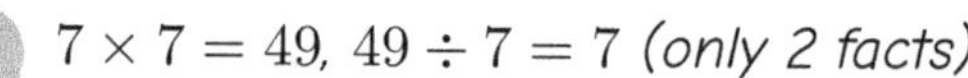 

1 $5 \times 9 = 45, 9 \times 5 = 45, 45 \div 5 = 9, 45 \div 9 = 5$　**2** $7 \times 7 = 49, 49 \div 7 = 7$ *(only 2 facts)*　**3** 72; *Yes*

4 35; *No — the correct answer is* $42 \div 7 = 6$　**5** 9　**6** 6

Section 3.9

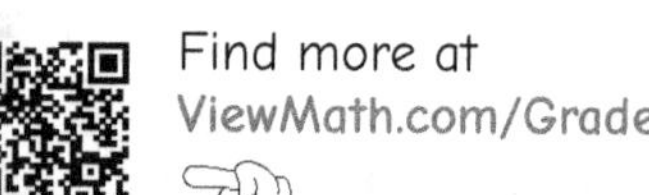
Find more at
ViewMath.com/Grade3

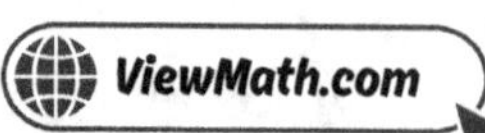

1 7 **2** 9 **3** 54 **4** 6 **5** 4 **6** 42 *stickers* ($n \div 6 = 7$, *so* $n = 42$)

> Section 3.10

1 45 *books* **2** 42 *miles* **3** 7 *students* **4** 7 *bracelets* **5** 24 *crayons* **6** 6 *boxes*

> Section 3.11

1 44 *crayons* **2** 32 *eggs* **3** 9 *students* **4** 6 *marbles* **5** 8 *flowers* **6** 9 *cars*

> Section 3.12

1 36; *multiply by* 4 **2** 70; *multiply by* 7 **3** 30, 36, 42; *add* 6 **4** 45, 54, 63; *add* 9

5 *Row* 4: 12, 28, 36. *Row* 6: 30, 54 **6** 18, 36, 54, 63, 81

> Section 3.13

1 68 **2** 90 **3** 104 **4** 91 **5** 156 **6** 120

📖 Chapter 4

> Section 4.1

1 $\frac{2}{4}$ **2** $\frac{3}{4}$ **3** 5 **4** 8 **5** $\frac{5}{6}$ **6** $\frac{3}{8}$ *of the pizza*

> **Section 4.2**

 $\frac{1}{3}$  $\frac{2}{4}$  6  *at 1*  0  $\frac{3}{4}$ *of the path*

> **Section 4.3**

 *Yes*  *No*  $\frac{1}{2}$  $\frac{1}{3}$  $\frac{1}{2}, \frac{1}{4}, \frac{1}{8}$ 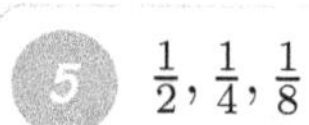 $\frac{1}{2}$ *of the ribbon*

> **Section 4.4**

 $\frac{2}{3}$  $\frac{3}{4}$  $\frac{4}{6}$ 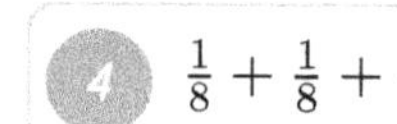 $\frac{1}{8} + \frac{1}{8} + \frac{1}{8}$  4  $\frac{2}{3}$ *of the bottle*

> **Section 4.5**

 2  4  4  *Yes*  6 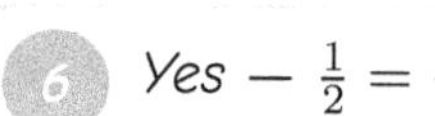 *Yes* $- \frac{1}{2} = \frac{2}{4}$

> **Section 4.6**

 4  1  0  6  5  $\frac{6}{6} = 1$ *whole pie*

> **Section 4.7**

 <  >  > 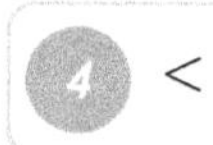 <  > 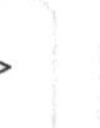 *Amy* $- \frac{5}{6} > \frac{3}{6}$

📖 Chapter 5

Find more at
ViewMath.com/Grade3

> **Section 5.1**

1. 4:00
2. 6:30
3. 10:15
4. 2:45
5. 8:27
6. 3:50

> **Section 5.2**

1. 2 hours and 30 minutes
2. 35 minutes
3. 3 hours
4. 7:45
5. 2 hours and 30 minutes
6. 1 hour and 30 minutes

> **Section 5.3**

1. 5 inches
2. $2\frac{1}{2}$ inches
3. $1\frac{3}{4}$ inches
4. $4\frac{1}{4}$ inches
5. $5\frac{1}{2}$ inches
6. 4 inches

> **Section 5.4**

1. grams
2. kilograms
3. 3,000 g
4. 4 kg
5. 5 kg
6. 8 kg

> **Section 5.5**

1. milliliters
2. liters
3. 2,000 mL
4. 5 L
5. 3 L
6. 6 L

> **Section 5.6**

1. 32 cents or $0.32
2. 40 cents or $0.40
3. 64 cents or $0.64
4. $0.45
5. $0.25
6. $0.95

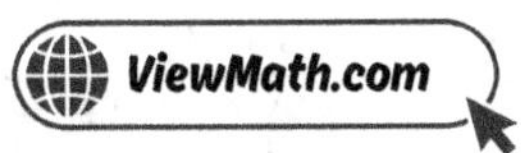

> Section 5.7

1 $3.75   2 $7.90   3 $5.25   4 $5.40   5 $3.75   6 $4.50

> Section 5.8

1 Need   2 Want   3 $10   4 $5   5 $2

6 No — she should buy the shoes (need) first. She has $6 left.

Chapter 6

> Section 6.1

1 12   2 6   3 Chocolate   4 4   5 36   6 15 stickers

> Section 6.2

1 8   2 Lily   3 5   4 11   5 22   6 7 students

> Section 6.3

1 6   2 6 inches   3 2   4 5   5 16   6 6 ribbons

Chapter 7

Find more at
ViewMath.com/Grade3

Section 7.1

1. Square
2. Trapezoid
3. Rhombus
4. True
5. False
6. Rectangle

Section 7.2

1. Cube
2. Sphere
3. Triangular prism
4. Cylinder
5. 4
6. 8 edges and 5 vertices

Section 7.3

1. 8 square units
2. 9 square units
3. 28 square units
4. 25 square units
5. True
6. 18 square feet

Section 7.4

1. 24 sq cm
2. 25 sq in
3. 36 sq ft
4. 28 square units
5. 16 sq cm
6. 24 square feet

Section 7.5

1. 16 cm
2. 26 in
3. 24 cm
4. 10 in
5. 12 cm
6. 26 meters

Section 7.6

1. $\frac{1}{3}$
2. $\frac{1}{4}$
3. $\frac{1}{6}$
4. 8
5. $\frac{1}{4}$
6. $\frac{2}{6}$

Section 7.7

 **1** Line segment  **2** Line  **3** Ray  **4** Acute  **5** Obtuse  **6** 2

⭐⭐⭐

Great job checking your work!

Keep practicing and you'll be a math star!

Find more at
ViewMath.com/Grade3

www.ingramcontent.com/pod-product-compliance
Lightning Source LLC
Chambersburg PA
CBHW060156120726
48004CB00007B/1564